Pelican Books
Pelican Geography and Environmental Studies
Editor: Peter Hall

Human Geography: Evolution or Revolution?

Professor Michael Chisholm has been Professor of Economic and Social Geography at the University of Bristol since 1972. Before taking up his present position, he was a staff member at Bristol, as well as previously at the University of Oxford, Bedford College, London, and the University of Ibadan, Nigeria. He has been a member of the Council of the Institute of British Geographers, and the Social Science Research Council, as well as currently being a member of the Local Government Boundary Commission for England.

Born in 1931, he was educated at St Christopher School, Letchworth, and St Catharine's College, Cambridge. In 1969 the Royal Geographical Society awarded him the Gill Memorial Prize. Among his publications are *Rural Settlement and Land Use* (1962), *Geography and Economics* (1966), and *Research in Human Geography* (1971). He was also co-author of *Freight Flows and Spatial Aspects of the British Economy* (1973) and *The Changing Pattern of Employment* (1973). He edited *Resources for Britain's Future* (1972) and jointly edited *Spatial Policy Problems of the British Economy* (1971), *Regional Forecasting* (1971), *Studies in Human Geography* (1973) and *Processes in Physical and Human Geography: Bristol Essays* (1975).

Human Geography: Evolution or Revolution?

Michael Chisholm

Penguin Books

Penguin Books Ltd,
Harmondsworth, Middlesex, England
Penguin Books Inc.,
7110 Ambassador Road, Baltimore, Maryland 21207, U.S.A.
Penguin Books Australia Ltd,
Ringwood, Victoria, Australia
Penguin Books Canada Ltd,
41 Steelcase Road West, Markham, Ontario, Canada
Penguin Books (N.Z.) Ltd,
182–190 Wairau Road, Auckland 10, New Zealand

First published 1975

Made and printed in Great Britain by
Richard Clay (The Chaucer Press), Ltd., Bungay, Suffolk
Set in Monotype Times

To Edith,
Annabel, Julia and Andrew

Contents

List of Figures

Preface

The immediate origin of this book throws light on both its structure and its purpose. The Social Science Research Council asked its various subject committees to prepare reviews of research in their respective fields. One of the main purposes was to make available, particularly to practitioners of other subjects, a clear but reasonably authoritative review of work being done. With the help of the Human Geography Committee members, I prepared *Research in Human Geography* (Heinemann, 1971). Shortly thereafter, I was asked by Penguin Books to write an expanded and modified version; the present book is the outcome. Needless to say, the process of rewriting and expansion has resulted in a metamorphosis, such that the present book bears only a faint resemblance to that written, with much assistance from others, for the S.S.R.C. I have tried to convey my own picture of how geography in general, and human geography in particular, has evolved to its present state and position. The result is a personal view of trends and developments, which though it is incomplete does, I hope, have some coherence, sufficient at least for the lay reader to find his way and for the practising geographer to put his own views into a recognizable perspective.

Writing this book has made me aware of the debt that I owe to innumerable people over a considerable period of time, a debt that has been contracted in countless direct and indirect ways. While it would be impossible to acknowledge everyone individually, I would like to express my particular thanks to some who, in various ways, have contributed to this volume. Oscar Backhouse started me on the geographical trail with his infectious enthusiasm, a trail that led me from St Christopher School to St Catharine's College, Cambridge. While there, Gus Caesar took me in hand in his inimitably warm but rigorous manner and showed me at least the rudiments of route-finding through the maze of academia. Thence to the Agricultural Economics Institute at Oxford, where working with Colin Clark proved to be a breathless chase, as he strode from peak to peak and I toiled along in the valleys below. Perhaps the other comparably formative experience was on the Social Science Research Council from 1967 to 1972, opening as it did my eyes to unsuspected views.

I also want to acknowledge the debt I owe to colleagues and to generations of students for the congenial and stimulating atmosphere at Bristol. For help with specific points I am most grateful to Malcolm Anderson, Peter Haggett, Les Hepple, Mike Morgan, Ron Peel and Frank Walker, all of the Geography Department at Bristol. Also, it was John Willats (Bradford-upon-Avon) who drew my attention to the coincidence in time of Renaissance interest in the problems of perspective drawing and the reawakening of curiosity concerning the mathematics of map projections. Professor Teruo Ishimizu, University of Nagoya, made numerous helpful suggestions.

Undoubtedly, though, my biggest debt is to my wife, Edith, and our three children, for their patience and forebearance. I only hope that the present book is worthy of them.

March 1974 MICHAEL CHISHOLM
Bristol

ACKNOWLEDGEMENTS

Grateful acknowledgement is made to the following for permission to reproduce material in this book.

To Edward Arnold (Publishers) Ltd, for the passage on pages 111–12 which first appeared in a paper 'Urbanisation, Domestic Planning Problems and Swedish Geographic Research', by A. R. Pred, in C. Board et al., *Progress in Geography*, vol. 5 (1973).

To Prentice-Hall Inc., for the passage on pages 92–3 which first appeared in Ronald Abler, John S. Adams and Peter Gould, *Spatial Organization. The Geographer's View of the World* (1971).

1. Introduction

Conceptual advances which have revolutionized the study of human geography as a science in the past decade.

Book advertisement, Wiley, 1973

The 'new' geography is no longer very new. It is over a decade since the so-called quantitative revolution took root in both the theory and practice of human geography.

Book review, *Times Higher Education Supplement*, 1973

When statements as divergent as the above two appear in the same year, albeit in different places and with different purposes in mind, both lay readers and practitioners of the subject may be forgiven if they feel that they have lost their bearings. Has something really important happened to human geography but has the flurry of innovation now subsided? In any case, in what ways, if any, is the subject identifiably different today from what it was one or two decades ago? And in what directions is development occurring at the present? It is with these and related questions that this book is concerned.

It would be as well to say initially something more about the purposes in mind when writing – what the book is intended to be as well as what it is not intended to be. In the first place, the account here presented must be regarded as highly personal; the reader who expects a carefully balanced discussion of all the major trends, as well as most of the minor ones, complete with judicious referencing to the full range of literature, will be disappointed. Perhaps the best way to express the standpoint adopted is as follows. When one is actually living through events, and indeed to some extent participating in them, it is difficult to take an entirely detached view and to weigh meticulously the significance of particular happenings, publications and individuals. Indeed, it would be invidious to pretend to do so. On the other hand, one cannot stumble blindly along in the comforting belief that because there is plenty of excitement we must be going the right way. As individuals, we must have some sense of direction and purpose and therefore a context in which to work. Perhaps, therefore, the primary purpose in writing is to convey an account of the direction and purpose of recent changes in human geography as conceived by someone fairly close to the scene. Whether this is coherent and appealing to others is a matter for the reader to judge.

Is there a subject that can properly be called 'human geography'? Much energy has been expended in attempts to define geography and its sub-divisions. However, it seems abundantly clear that other subjects have an equal difficulty, and in an important sense time spent on subject classifications is time lost for the examination of substantive problems. Therefore, in the present context it does not seem useful to dwell on a problem which, virtually by definition, lacks an answer that will be universally acceptable. On the other hand, some definition must be given, at least as a starting point, so that we may avoid the mistake Hartshorne (1939) made (see Chapter 2). He set out to establish the nature of geography by an examination of published work, ostensibly eschewing any prior definition of the subject. At no point does he explain the basis on which work was classed as 'geographical' or not, yet that classification must largely determine the nature of the subject identified.

As a rough-and-ready definition, let geography be taken to cover three related themes:

1. The recording and description of phenomena at or near the surface of the earth (the literal meaning of the word geography).
2. The study of the interrelationships of phenomena in specified localities.
3. The examination of problems which have a spatial (terrestrial) dimension, especially to identify the significance of space as a variable.

Within this definition, human geography may be distinguished from physical geography according as the main emphasis is on human phenomena or phenomena of the natural environment. Necessarily, there is a middle ground where the two are given approximately equal attention, and indeed some writers hold that this middle ground is precisely what geography is all about:

I wish to press the general proposition that, as an academic group, we geographers face the choice between extinction on the one hand, and, on the other, integrated work of the kind which our predecessors claimed to be particularly suited to geography and geographers.

(Dury, 1970, p. 31)

If Dury's view is accepted, it is evident that a geographer who develops a specialized approach is in danger of becoming the practitioner of another discipline. This implies that there is an 'ideal' geographer and that we ought all to approximate to this ideal. Such a proposition is in fundamental opposition to the view taken by the present writer, and implied in the preceding paragraphs, namely, that one subject is distinguished from another as coalitions of individuals for which the within-group variance of interests is less than the between-group variance. Subjects may therefore be conceived as analogous to the clusters of phenomena that are obtained from formal procedures for partitioning multi-variate data sets (Chapter 3).

Ultimately the form of the question or questions which distinguish geography from the other sciences is not too important. No other science consistently concerns itself with distributions of phenomena in terrestrial space; no other science consistently concerns itself with spatial structure. The questions about location, spatial structure, and spatial process which we ask and answer distinguish geography from the other sciences.

(Abler, Adams and Gould, 1971, p. 61)

How long a historical perspective?

In the discussion of how a subject evolves, and in an attempt to examine some of the more important ideas which are current currency, it is not very meaningful to establish an arbitrary time period and confine attention to work carried out in it. The more realistic strategy is to write history 'backwards', in the sense of focusing attention on the present and extending one's gaze backwards in time only so far as is necessary for the purpose in hand. Thus, in some contexts it is relevant to refer to events only since the last war; elsewhere, a reasonable perspective can only be obtained if the retrospective view extends over several centuries.

Although it is undoubtedly to the German geographers of the

nineteenth century that we owe the formalization of geographical work into the discipline we now call geography, and though early in the present century it was the French who probably led the field, the present book concentrates on the English-speaking world. Effectively, this means that pride of place is given to literature published in Britain and North America, the greater part of which is twentieth century and indeed mainly published since the First World War.

Scale and growth of geography

The present book is intended as a discussion of ideas, linked with individual authors and especially with those persons who can be thought to have made an outstanding contribution to our subject. Such a discussion provides no concept of the scale of resources devoted to the study of geography. Some brief comments on this topic will therefore set the scene for what follows (see Stoddart, 1967). Note that in the following paragraphs no distinction is drawn between physical and human geography and that for the present purpose it is the *whole* subject with which we are dealing. Also, the nature of the information available means that only rough-and-ready indicators of growth and scale can be employed.

Figure 1 shows that in 1750 there were two scientific geographical journals published in the world and that up to 1800 the rate of growth was substantially greater than for all scholarly periodicals. From 1800 to 1950, world-wide publication of new geographical journals led to a doubling every thirty years, a growth rate one-half of the rate for scientific periodicals generally. However, between 1960 and 1971 there has been an acceleration in the output of new geographical journals, the total number founded rising from about 1,600 to about 2,400 in the period (*Area*, Vol. 5, 1973, p. 81). At this rate, the number will double in approximately twenty years, though whether this increased tempo of activity will be maintained remains to be seen.

The world-wide growth in geographical societies is equally interesting (Figure 2). From the establishment of the Société de Géographie de Paris in 1821, there was a startling diffusion of societies around the world until the end of the nineteenth century. Thereafter, the expansion has been less hectic, with a doubling every forty years to bring the total number to the two hundred mark early in the second half of the twentieth century.

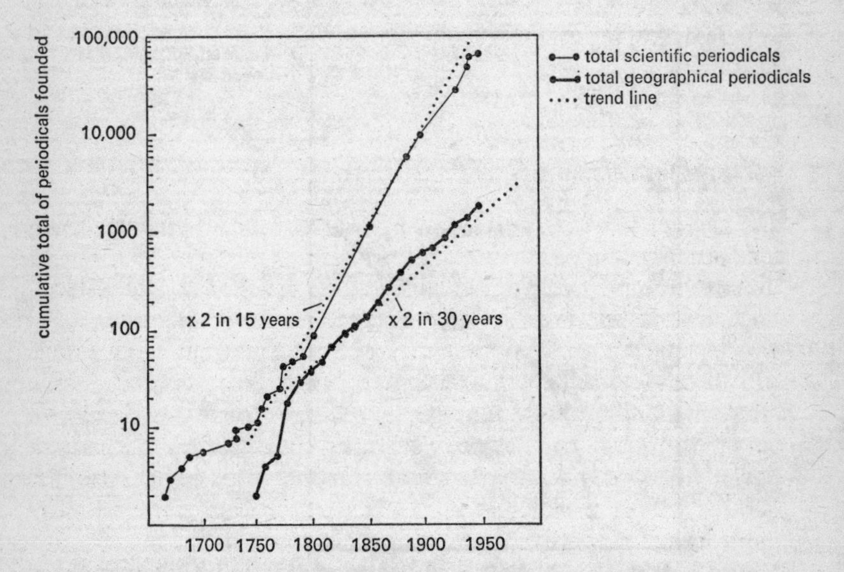

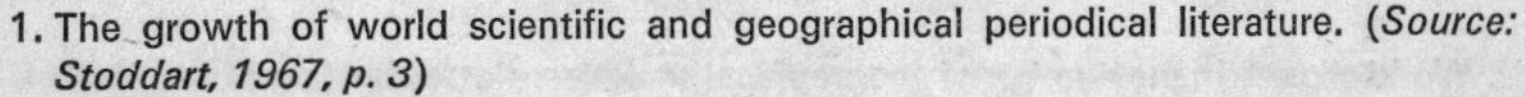
1. The growth of world scientific and geographical periodical literature. (*Source: Stoddart, 1967, p. 3*)

A third indication of the scale of geography is the fact that in 1964 there were 4,400 professional geographers in the world and that the ‘output’ of higher degrees in the subject represents a doubling of the number of ‘trained’ geographers in Britain and the United States in a period of the order of eight years (Stoddart, 1967, pp. 5–6). In terms of higher degrees awarded, geography represents 0·5–1·0 per cent of total research effort.

With exponential growth apparently the order of the day for the number of professional scientists, of professional societies and of publications both in geography and for all sciences, Stoddart is undoubtedly right to comment:

> The crisis in education lies in the question whether the rate of downward diffusion of this rapidly expanding body of knowledge increases to keep pace with it, or whether the efficiency of filters at different levels in the education hierarchy can accommodate the exponentially increasing flow of information reaching them.
>
> (Stoddart, 1967, p. 6)

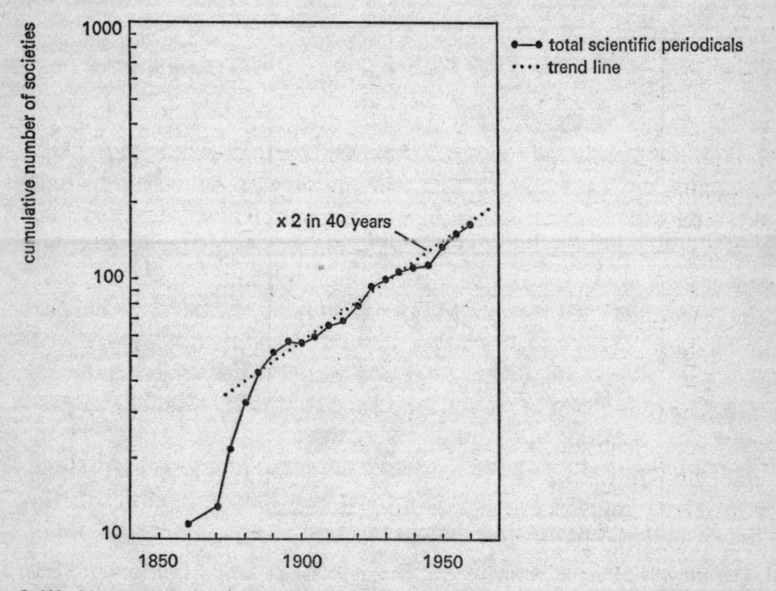

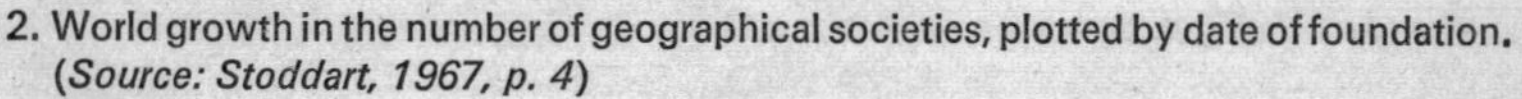

2. World growth in the number of geographical societies, plotted by date of foundation. (*Source: Stoddart, 1967, p. 4*)

The present book is an attempt to provide just such a filter, in the hope that it will be of particular assistance to students and the interested lay reader. Yet, the reader may well ask, is there not virtually a library of such books already available and therefore what can be the justification in attempting to squeeze yet another volume onto the shelves? The fundamental answer is that, so far as the present author is aware, nobody has adopted the approach embodied in the present text and it is therefore a reasonable hope that readers will find a view of the subject that will at least be interesting; hopefully it will also be persuasive.

2. Origins

There is a history in all men's lives,
Figuring the nature of the times deceas'd,
The which observ'd, a man may prophesy,
With a near aim, of the main chance of things
As yet not come to life, which in their seeds
And weak beginnings lie intreasured.

William Shakespeare, *King Henry IV, Pt 2*

Though the past is only a partial key to the present and future, it is impossible to understand the contemporary scene without some knowledge of its origins and development. Thus, the questions geographers ask and the way we go about providing answers are in considerable measure conditioned by the traditions to which we are heir. It is only if one has some understanding of this history that one can begin to answer the question posed at the start of Chapter 1 on the nature of the changes that have occurred in the subject in recent years. However, where should one start and with what detail proceed?

An analogy will help to provide the answer to this question. Suppose we had the task of describing in words the nature of the various lands on planet earth so as to build up a complete picture of the globe. We could start at any point and traverse the planet along any one of an infinite number of circles. When one such traverse had been completed, we would want to choose another starting point and another circular route. After a few such circumferential tours, the general configuration of the earth would begin to be apparent. If we were sufficiently persistent, a complete description of the world could be compiled in this way. In the process, the first few traverses might well produce accounts in which it was hard to relate information about one place to knowledge of other locations. As the process continued, it would be found that the circular tours interlock and so the mutual relationships of the parts would become increasingly apparent. Once the main outline had been recognized, particular areas could be explored more thoroughly and also the full significance of fine detail appreciated.

The present chapter consists of several such metaphorical girdles thrown around the subject geography, designed to show how the subject evolved and the nature of the questions geographers pose. Rather than take a strictly chronological approach, several significant themes have been identified. The process of discovery and exploration is described, as well as the associated problems encountered in recording geographical data. Hence follows the question of what kinds of data are significant, something that depends on the type of scientific explanation at which one aims. However, the very notion of explanation involves philosophical problems of a high order and philosophical concepts have themselves evolved with the passage of time. In particular, scientists have become impatient with inferences about causal processes and have sought to measure these processes directly. As a consequence, geographers have been led to adopt the modern 'scientific' approach, including the use of quantitative techniques. These techniques are appropriate for the analysis of certain kinds of problem, but the question does arise whether they are applicable to all kinds of situation, or whether in fact other approaches are necessary in at least some cases. Thus, in this chapter I attempt to provide a synoptic view of major trends in geographical thought as the basis on which to pursue in subsequent chapters some of the more important themes.

Referring to the development of geography in western Europe, Abler, Adams and Gould (1971) recognize three major phases in the period since the Portuguese initiated the great Age of Discovery in the fifteenth century. These they identify in terms of the pre-eminent problem, the overriding issue that attracted the labour of scholars. The dates assigned are necessarily arbitrary and the periods do of course merge into each other; nevertheless, the shifts of interest are readily identifiable. Until 1800, the three authors state, the overriding problem was the development of a metric for recording absolute locations, the evolution of map projections and the associated task of compiling and revising maps. In considerable measure, the advancement of geography was synonymous with the development of cartography. From about 1800 the emphasis shifted to the problems posed in recording what places, now accurately located, are like and how far one place is similar to or different from another. The classical works of Alexander von Humboldt and Carl Ritter set the scene for the scientific observation and recording of phenomena in terms that have a manifest continuity to the present day. The third period is identified as starting about 1950 and is characterized by attention to relative, as against absolute, locations. The authors conceive of absolute location as being the use of longitude and latitude to give a unique co-ordinate reference to every place, from which the geometrical distance and azimuth to any other place can be determined. Relative location, in contrast, is the transformation of geographical co-ordinates into a metric of cost, time, etc. What was previously fixed and unchanging – at least within the range of human sensibilities unaided by modern methods of measurement – is now conceived as a constantly varying set of relative locations. While there can be no doubt that this last change in outlook is both real and important, it is debatable whether it really is 'the most fundamental change in the history of geography as it opens an almost

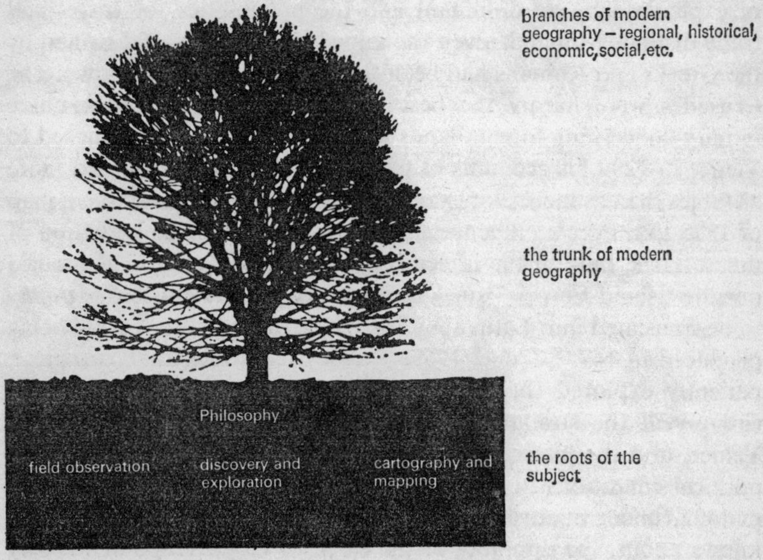

3. Geography's tree of life.

infinite number of new worlds to explore and map' (Abler, Adams and Gould, 1971, pp. 72–3).

In any case, the threads that make up modern geography are considerably more numerous than is indicated above, though closely related. If we visualize modern geography as a tree, it may be represented as in Figure 3. The three main roots are indicated as the process of discovery and exploration of the world, map-making and field observation. These roots do not connect directly to the trunk but are mediated through philosophical notions concerning the nature of reality. Since philosophical concepts have changed over time, the significance of the three roots must periodically be re-examined and re-interpreted; by the same token, our conception of the trunk and branches has evolved over time.

EXPLORING THE WORLD

Medieval Europe was a curiously cloistered place. Theological modes of explanation were dominant and the possibilities for travel and trade distinctly limited. Even the knowledge of the world gained by the Greeks and Romans had been 'lost', lying largely unknown. The Crusades were hardly the best way of conducting dispassionate inquiry concerning foreign lands and when Marco Polo returned to Venice in 1299 his accounts of China and Asia were regarded more as fables than as facts. For example, it was not until the Catalan map of 1375 that Polo's influence can be detected in the construction of maps. Thus, only in the fifteenth century had interest in the world outside 'island Europe' quickened enough for Ptolemy's *Geography* to be translated into Latin about 1406, the first printed version being produced in 1475 (Crone, 1968, p. 65). Meanwhile, the Portuguese patiently explored the African coast for a sea route that would circumvent the stranglehold on trade with the Far East held by Venice, and the interruptions to that trade occasioned by unstable political conditions in Asia. While Columbus's voyage to America in 1492 (under Spanish patronage) is something every school child knows about, the doubling of the Cape of Good Hope in 1497 by Vasco da Gama was equally significant. If not to emulate Puck and girdle the earth in forty minutes, the way was paved for Man's first circumnavigation of the globe by Magellan (1519–22).

Thus was phantasy converted to fact. The equatorial zone was shown not to be too hot for human habitation, mythical monsters were not encountered at every turn, even though there were many deadly perils. But as one problem was resolved, others came to the fore; as the earth was accepted to be a sphere (notwithstanding that a Flat Earth Society still exists), it was deduced that there must be a large continental mass in the southern hemisphere else the world would tip over. Erroneous though this deduction was, it provided one of the stimuli for the exploration of the Pacific Ocean: the main outlines of Australasia were known by the end of the eighteenth century, but the first land of Antarctica was not sighted until 1839, sixty-six years after Cook suggested the existence of that continent. With the exception of Antarctica, which still presented problems concerning the extent of land even as late as the 1930s (Baker, 1945, p. 486), all the main continental outlines were established with reasonable accuracy by about 1800, representing the culmination of some 400 years' endeavour. The interiors, of course, remained mysterious for much longer. Where did the river Nile have its source, and what happens to the headwaters of what we now call the river Niger as they flow northwards to the Sahara Desert? Altogether, therefore, the epic era of exploration continued through the nineteenth century at least until the tragic death in 1912 of Scott, returning from the South Pole after being beaten to the post by Amundsen.

Viewed from the 1970s, when mankind has set foot on the moon several times, and terrestrial explorers are equipped with radios and supported by air-rescue facilities, it is easy to forget the excitement and intense interest that attended Man's discovery of his own planet. More important, it is all too easy to forget just how recent this has all been and that, for example, people alive today knew Scott and his companions. Thus, the major root of geography has experienced vigorous growth right into the present century.

Early in the process of discovery and exploration, the need for efficient methods of storing and retrieving information about the world became pressingly urgent, not least for reasons of commerce. The force of this point is aptly illustrated by an episode in the life of

Galileo. When he developed an improved telescope he made his fame – with the merchants of Venice, who could now identify homeward-bound ships while they were still several hours' sailing from port. Only subsequently did Galileo turn the telescope to the heavens for purely scientific purposes. In much the same way, maps and other geographical records were of great commercial interest for the fortunes that might be made. Such knowledge also had a clear military value, as indeed is true to the present day. Purely scientific interest in geographical information came later, as it were in the wake of the other two concerns. Thus it was that the first intellectual challenge in geographical work was to find efficient data-storage techniques.

RECORDING DATA

1. Map construction

The ancient Greeks visualized the earth as a sphere and the experiment by Eratosthenes (b. 276, d. 192 B.C.) to determine the size of the globe is rightly famous. Ptolemy's world map was an attempt to reproduce this sphere in two dimensions and for this purpose lines of longitude and latitude are employed. But this method for transforming data into map form was 'lost' in the Middle Ages. Thus, in the thirteenth and fourteenth centuries another and inherently inferior system was evolved, known as the 'portolan charts', culminating in the Catalan world map of 1375. The charts proper depicted a relatively small section of the globe – the Mediterranean and adjacent regions – were constructed as aids to navigation and were indeed reasonably accurate. On the assumption that the earth was flat, they were constructed by the intersection of compass bearings, that is, by a version of trigonometry recognizable as the basis for modern mapping. However, given the assumption of a flat earth, the system could only be applied to small areas with any real precision. With the rediscovery of Ptolemy's work in the early fifteenth century and the construction of Martin Behaim's globe in 1490, the way was opened for major cartographic advances, for the accurate recording of where places are in terms of the system provided by longitude and latitude, and hence for the construction of charts useful for navigational purposes over the whole world (Crone, 1968; Skelton, ed., 1964).

The fundamental problem with which cartographers wrestled was the fact that on a globe it is possible to portray three basic properties of the world's surface features – shape, area and distance – as well as the correct bearings between pairs of places as a derived property. In a two-dimensional framework – a map – these four attributes cannot be retained simultaneously; indeed, a decision must be made as to which property or properties are required, and in what degree. For example, Mercator's projection, which appeared posthumously in 1569, is so designed that a line of constant bearing is represented as a straight line, a quality of enormous value to navigators but one that introduces great distortions of shape, scale and area toward the poles.

A fascinating aspect of the rediscovery of Ptolemy in the early fifteenth century and the subsequent development of modern map projections is its coincidence in time with the efforts of painters and architects to solve an identical problem. Giotto completed his magnificent frescoes at Assisi about 1300, using a rather crude technique in perspective but without having discovered the formal rules for portraying three-dimensional objects as two-dimensional paintings. These rules were not formally codified until Alberti published his book on the subject in 1436. Thus, cartographers were wrestling with a special part of a problem which had very general applications, and was a fundamental ingredient of the development of modern scientific thinking.

Advance in the construction of map projections continued after Mercator. Various cylindrical projections were devised with properties other than the one of constant azimuth, including the property of equal area, and conical projections were also constructed. By 1805, when Mollweide invented his equal-area projection for the world, techniques were adequate for representing the whole globe in map form; though improvements have been made subsequently, and new projections invented, the main innovative phase had ended. With the perfection of projections and survey techniques, it became logical to establish national surveys. First in the field was the British Ordnance Survey, established in 1791 and publishing its first map in 1801 – of Kent and adjacent areas, a significant choice in the Napoleonic wars. Other national surveys were quickly established, and thus the nineteenth century dawned with both the technology for accurate mapping and the beginnings of a proper administration for the purpose.

Three final points deserve notice at this juncture. In a very real sense, the roots of quantification in geography lie not in modern statistics but in the art and science of map-making. Admittedly, it was a different kind of quantification, related to a determinate geometry used for accurate description, and not, as in the modern era of quantification, designed for explanatory purposes in a framework of probability. In the second place, Abler, Adams and Gould are surely mistaken to describe the mapping phase as a concern with absolute location. It would be truer to say that the great cartographers were primarily concerned to portray the relationships between places in an objective manner, that is, in a manner that allowed of precise repetition given that certain rules were followed, but that nevertheless the locations are all relative (Blaut, 1961). Leading on from this, however, the most important single feature of the interest in cartography is the problem of data transformation; transformation from three dimensions to two is merely a particular problem in a much wider category of transformations with which geographers have been concerned.

2. Phenomena in place

It is all very well to record accurately the locations of capes and bays, rivers and towns, but what are these places *like*? For their systematic attempts to answer that question, two names stand out above all others, Alexander von Humboldt and Carl Ritter (Dickinson, 1969). Humboldt, trained as a mining engineer, embarked on a career as natural scientist and explorer when he inherited an income in 1797. He spent five years travelling in Latin America and is famous for the meticulous care of his observations and the way in which he recorded and related his findings concerning the distributions of rocks, flora, etc. In particular, he was concerned wherever possible with measurement, carrying with him forty different kinds of instrument, ranging from telescopes to barometers. Though he published various works earlier, it was not until 1845 that the first two volumes of his *Cosmos* were published (it ran to a fifth but incomplete volume).

Carl Ritter could fairly be described as an 'arm-chair' geographer. He set out to compile a geography of the world, arranging his material on a continental/regional basis. The first volume of *Erdkunde*, this encylopedic work, was published in 1817, followed by eighteen more volumes at intervals, but even then the task had not been completed by the time of his death. Ritter worked in libraries, culling his information from the records of explorers and travellers but making every effort to check the veracity of his facts. He wrote within the framework of an elementary classification of terrain based on relief – mountains, lowlands, etc.:

> This classification, summed up in the introduction to the *Erdkunde*, formed the basis of Ritter's treatment of the continents. Superficial as the scheme may be, it afforded a new regional method of description which was distinct from the usual method of dealing with political units. In each division the main features of the relief and drainage are described, followed by climate, major products, and population.
>
> (Dickinson, 1969, p. 41)

German geographers in particular subsequently devoted a great deal of effort to refining the framework within which to describe the collections of phenomena that occur in specified places. Perhaps the single most important figure to mention is von Richthofen, who, having travelled extensively in China, returned to Germany in 1872 and subsequently developed what has become the classical approach to regional geography – the framework that starts with the natural environment and proceeds to the distribution of man and his activities. This tradition was carried to its finest achievements by French scholars: Reclus published the twelve-volume *Nouvelle Géographie Universelle* (1876–87), though the more famous edition was published in the twentieth century by Vidal de la Blache, who also published several other notable regional volumes early in the present century (e.g., 1903).

3. Thematic maps

The problem with which these and many other workers were struggling can be related to Berry's (1964) concept of a geographical data matrix (Figure 4). He envisaged an array of locations, for each of which there is a set of 'facts'. Thus, any particular fact can be

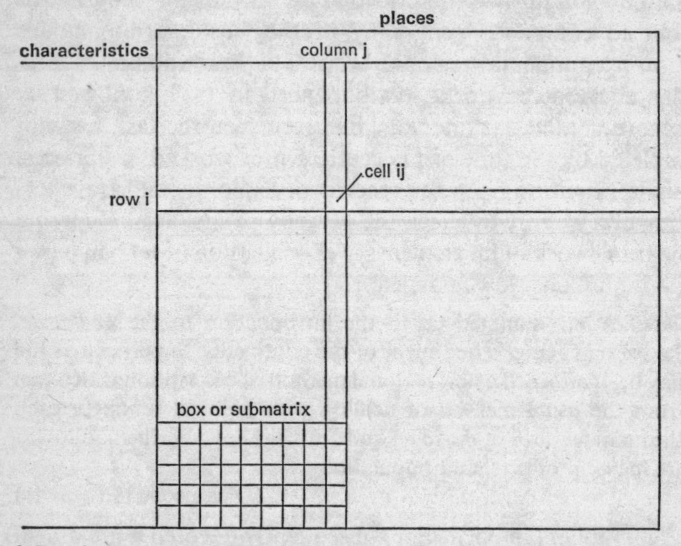

4. Berry's geographical data matrix. (*Source: Berry, 1964, p. 6*)

located by reference to its geographical location – represented by the columns – and its classification by kind of information (rows). A fundamental problem with which nineteenth-century geographers wrestled was the way in which to arrange the rows, that is, how they should be grouped and in what sequence to be studied. The nature of the basic scheme which they evolved is shown in Figure 5, where the various characteristics or attributes are grouped into broad headings that can be treated sequentially from the top (geology) downwards for any or all of the columns, which represent places of larger or smaller extent. The order in which the columns were placed was not regarded as a major issue.

However, as more and more information became available, albeit imperfect, an increasing number of the cells in the matrix were filled. Consequently, it became natural to ask whether the arrangement of the columns might be significant in considering any one class of data, that is, whether the rows of the matrix might yield interesting information concerning the spatial distributions of particular phenomena. One of the pioneering compilations in this vein was the Berghaus atlas published in 1837 (Freeman, 1967, p. 59), with maps

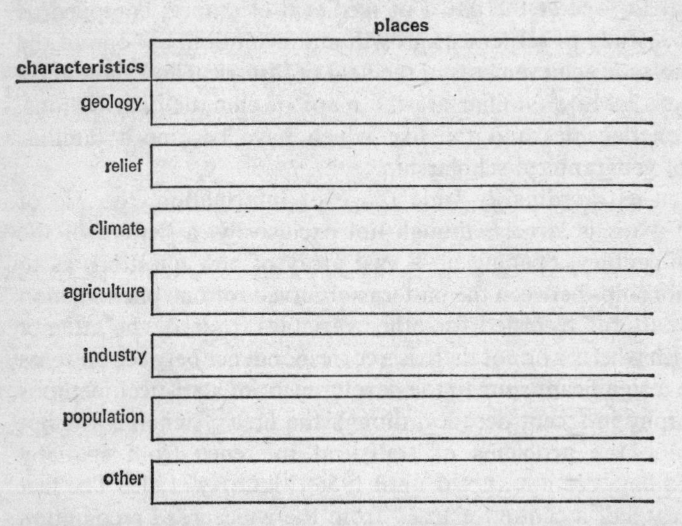

5. The geographical data matrix as a framework for regional geography. (*Source: adapted from Berry, 1964*)

showing the relationship between climate and vegetation. Since data were scarce, many of the early maps were impressionistic and they were in any case subject to revision as new information became available (Wilkinson, 1951). However, as the present century approached, the scene was set for Köppen's first version of the world's climatic regions, published in 1900 (Freeman, 1961, p. 315). His regions were largely based on plants as indicators of climatic conditions, rather than upon direct observation of temperature, precipitation and other climatic parameters. If not the first attempt to define the world's climatic regions, this was by far the best to date, had an immediate impact on his contemporaries and has remained influential to the present day (Hare, 1951). In particular, it was only five years later, in 1905, that Herbertson, much influenced by Köppen, published his famous paper on the natural regions of the world. From about 1900 onwards, thematic maps became part of the geographical stock-in-trade, not only for world-wide distributions but also at the regional and local scales. For example, Dickinson (1969, p. 128) regards Schlüter's reconstruction of the forested area of central Europe at the dawn of medieval clearance, compiled as part of his study of settlement growth and evolution, as 'one of the great scholastic achievements in the field of historical and geographical analysis'. More familiar are the maps of climatic distributions, population densities and the like which have become a familiar feature of geographical scholarship.

Over an extraordinarily wide range of information, the use of thematic maps is largely, though not exclusively, a feature of the twentieth century, opening up a vast array of new questions as to the relationships between the patterns observed for one phenomenon and the patterns recorded for other variables. Indeed, the attempt to establish whether or not there is a correspondence between patterns has been a significant spur to the development of statistical methods in geography in recent decades, though the first systematic attempt to examine the problems of statistical inference from spatially organized data did not appear until 1956 (McCarty et al.). The fact that nearly half a century elapsed from the widespread recognition of spatial patterns in the distribution of particular phenomena to serious attempts at rigorous statistical testing of spatial associations may seem surprising. The explanation lies partly in the need to await developments in inferential statistics generally and partly in geography's retreat from the search for explanatory models occasioned by disillusion with environmental determinism (see p. 37).

WHAT KINDS OF DATA?

Given the evolution of geography from the days of exploration and the scientific recording of information in the field, the subject is heir to two related traditions, one explicit and the other implicit. The explicit tradition is of field work, the gathering of information by the researcher himself, whether by recording such natural phenomena as angles of slope and the movement of glaciers or such human activities as patterns of land use and the spatial distribution of dwelling types. Very often, this field work was closely associated with the need to carry out plane table, chain or other surveys, so that the locations of phenomena could be precisely fixed. It is probably true to say that until the 1950s, at least in British universities, surveying/map projections/cartography held the central position as necessary skills for the geographer which is now held by statistics. It is certainly true that field work remains an essential part of much geographical work; Wooldridge and East (1966) rightly devote a chapter of *The Spirit and Purpose of Geography* to it.

Much more important, though, is the implicit habit that became widespread in geographical work in the nineteenth century and remained at least through the first half of the twentieth century. With field observation, one can record 'things', either visible phenomena such as mountains and villages or invisible ones that are tangible to the senses and susceptible to measurement, such as temperature. Therefore, it is with concrete phenomena, visible facts on the surface of the earth, that substantive work has been principally concerned. When Hartshorne and others write of the interrelationships of phenomena in place it is normally implicit that these are visible phenomena or ones that are measurable. A very simple example will illustrate what is meant by this statement and the fallacies to which this mode of thinking can lead. To examine the reasons for patterns of agricultural production in Britain, it is clearly essential to think of the relationship between particular crops and livestock on the one hand and conditions of the natural environment, such as climate and relief, on the other. However, to limit the discussion to these tangible variables would clearly be inadequate. Even to note where the main urban markets are located and the costs of transport to them would add only a limited amount to our understanding. An adequate study must also take account of government policies – subsidies, tariffs, etc. – and the pattern of competition from abroad; neither of these parameters is visible or measurable in the field within Britain. On the other hand, it is quite possible that either or both of these considerations outweighs the relationships that can be directly measured. As a broad generalization, it is only since the 1950s that this fundamental point has gained widespread, though by no means universal, acceptance amongst geographers (see, for example, James, 1971, pp. 79–80). Suffice it at this stage to recognize the point and note that it ties in very closely with the problems of the inferential steps required to explain patterns that have been observed (p. 147). Thus, in the inter-war years there was 'a growing belief among the rising generation of students . . . that the geography we were being taught was proffering bogus explanations' (Fisher, 1970, p. 382).

ATTEMPTS AT UNDERSTANDING: FROM REGIONS TO SYSTEMS

The nineteenth century still carried a strong impress of religious attitudes, one of which in particular was incorporated into a great deal of geographical work. It was widely believed that Man had very little control over his destiny, except in so far as a state of grace could be attained by religious observance and leading a good life. Particularly relevant in the present context is the notion that, though God could not be known in a direct sense, He is present everywhere and all of nature represents His handiwork. On this basis, scientists and other thinking persons attempted to reconcile science and religion in the following manner. The greater our understanding of nature, the greater our knowledge of the works of God and therefore the nearer we would approach Him. The formal term for this view of the world is 'teleological', that is, 'dealing with design or purpose, especially in natural phenomena' (*Shorter Oxford English Dictionary*).

An extreme exponent of this viewpoint was Ellen Semple (1911), who personified nature as Mother Nature who gave Man 'commands' regarding the activities he might or might not carry on and where, the clothes he should wear and the food he should eat, etc. Semple consequently adopted an extreme position regarding environmental determinism, of which more is said elsewhere (p. 37).

More important in the present context is the concept of the

'oneness' or 'wholeness' of things. This view found special expression in the writings of the German philosopher G. W. F. Hegel (1770–1831). Hegel's philosophical attitudes amounted to 'an attempt to comprehend the entire universe . . . to know the infinite and to see all things in God' (*Encyclopaedia Britannica*, 1971, vol. 11, pp. 300–301). Although Hegel himself was a man of many ideas, and though German philosophy was extraordinary for its quality and variety, the ideas embodied in the above quotation have had a curiously profound influence upon geography. This influence has been largely obscured by the, perhaps undue, attention geographers have given to Immanuel Kant – the only modern philosopher of major stature specifically to place geography on the map of learning (May, 1970): 'Kant, the great German master of logical thought, gave geography its place in the over-all framework of organized, objective knowledge' (James, 1971, p. 13). Among Kant's many ideas, German geographers were especially attracted to the concept of the uniqueness of phenomena and the absence of any basis for formulating general laws about events (Schaefer, 1953). Unique events cannot be predicted, but they may be understood as part of some wider totality. Thus, in the name of Kant an essentially Hegelian (holistic) view of the world was adopted, especially by Hettner, who in turn exerted a major influence upon Hartshorne (*The Nature of Geography*, 1939). Although not an orthodox view, I think it is fair to say that though Kant is widely quoted by geographers, and revered for giving us a place in the sun, it is really the ideas of Hegel and his concept of the wholeness or totality of things that has been pervasive in geographical work. This in turn has blinded geographers to the distinction, clearly made by Kant, between things metaphysical that are matters of faith and things that are amenable to scientific analysis, a distinction that does not sit easily with Hegel's philosophical viewpoint.

The geographer's most important traditional form of this concept of the wholeness of phenomena is the concept of region or landscape – the meaning of the German term *Landschaft* is not fully caught by the English equivalent. By taking a defined and perhaps relatively small area, the geographer could attempt to understand the totality of phenomena and their interrelationships. Furthermore, the notion of landscape, embodying the sum of natural and human phenomena, was thought to enable the student to make teleological inferences concerning man and his relationship to the natural environment. This train of reasoning led quite naturally to the problem of how to define regions and indeed to the proposition that regions exist as objective realities, independent of the observer (Hartshorne, 1939; Wooldridge and East, 1966; Minshull, 1967). This is an idea that has a direct parallel with the Marxist notion of 'objective social conditions', such conditions being conceived as existing independently of the viewpoint of the social commentator, reformer or revolutionary (for instance, Harvey, 1973, p. 129). Indeed, the idea was widely canvassed that regions provide the 'objects' that geographers study, thereby giving a basis on which to define the subject.

To study the totality of phenomena is indeed a worthy ideal; the only thing wrong with it is the magnitude of the task, such that only a very few men have the ability to approach anywhere near the goal that is set (Wooldridge, 1950). In overt terms, this difficulty became manifest with the failure of scholars to agree on a comprehensive set of regions as the framework for geographical studies. An extreme example of this difficulty is provided by Sinnhuber (1954). He reviewed the then extant literature on central Europe and compiled a fascinating map showing the conceptions of sixteen authors concerning the region's boundaries. With the exception of Austria and western Czechoslovakia, which all authors agreed were in central Europe, the remarkable feature of the map is lack of agreement over an area extending from northern Italy to southern Sweden and from the Netherlands to the Ukraine. The disenchantment with regions has been such that whereas regional accounts and regionally based analyses formed a very large share of geographical writing – at least of books – up until the 1950s, regional studies are now regarded as somewhat *passé*. With the outstanding exception of Brookfield's (1971) analysis of Melanesia, regional geography is not now seen as the major focus for intellectual effort (see Fisher's 1970 lament), even though James (1971, p. 78) and others are of opinion that 'the regional concept constitutes the core of geography'.

If regional geography is thus in some disarray, partly owing to difficulty experienced in agreeing on the network of regions, the desire to find a framework within which to examine 'wholes' rather than 'parts' has led some scholars to advocate the advantages of a systems approach (Stoddart, 1965; Haggett, 1965; Chorley and Kennedy, 1971; Berry, 1973; Warntz, 1973; Langton, 1972). Haggett has summarized the major classes of system. The fundamental proposition is that:

> Relations between individual components are built up by statistical association to produce positive or negative bonds. Changes in the level of one component cause associated changes in other components. Such systems vary in the number of components, the strength of the links between them, and the arrangement of the links into positive or negative feedback links.
>
> (Haggett, 1972, p. 44)

Systems can be viewed in a hierarchical sense, since the output from one system may be an input to another. For example, attempts to develop operational models of cities are based on linking together sub-models of land use, transport, housing, etc., each such sub-model being in itself a system (McLoughlin, 1969).

In geographical studies, all systems are open in some degree, that is, there is transfer of either mass or energy, or both, across the system's boundaries. The relative importance of these transfers will clearly vary from case to case, and will also be affected by the purpose for which the system is being defined. Thus, there will always be a spectrum of systems, ranging from those that are readily identifiable to those which are susceptible to much dispute. Furthermore, a geographical region is but a special kind of system. These points are readily perceived from the following tabulation.

	REGION	*OTHER SYSTEMS*
Definition generally agreed	River drainage basin	Central heating system
Definition disputed	City region	Political decision system

A further feature of the fashionable interest in systems is the continuing implication that regularities result from 'purposive' behaviour of the elements in the respective systems with the teleological inference that there must be some greater 'purpose' external to the system in question (Berry, 1973; see the interesting discussion in

Buckley, 1968). Altogether it seems likely that by the end of the present decade it will be generally accepted by geographers that while systems, like regions, provide a useful framework within which to work, they are all too frequently intangible things that with maddening regularity retreat from the researcher – just as the bag of gold at the rainbow's end eludes the seeker after riches.

In this context, a clear distinction must be made between the scientific ideal to which we strive and the practical realities of what is feasible for most workers. Colin Clark once remarked to the present writer, green from graduation, that 'the best is the enemy of the good'. In other words, clearly the best work comprehends the whole of the relevant totality – region or system – but in any generation it is given to only one or two men to achieve the superlative. Thus, while an integrated approach to a 'whole' problem is clearly the right aspiration, it would be wrong to be downcast because few of us are capable of achieving it. After all, the difficulty is a universal one, applying to all academic disciplines as well as our individual private concerns.

THE SEARCH FOR EXPLANATION

Although the tradition that knowledge equals understanding is a long one, and had for many centuries seemed a sufficient justification for scientific effort, *the* great intellectual revolution of post-medieval times was the rational search for explanation and its concomitant, prediction. From the seventeenth century onwards, the pace of scientific discovery quickened, old problems were solved and new marvels unearthed. This scientific awakening led to two publications that have had a widespread influence on the way geography developed. *The Principles of Geology* by Sir Charles Lyell was published in three volumes, 1830–33, and Darwin's *On the Origin of Species* appeared in 1859. The former established the evolutionary processes by which the surface of the earth is moulded, on the principle of irreversible change followed by yet further irreversible change. Lyell's work was predicated on the assumption that geomorphological processes seen operating today are the same as those that were moulding the earth's surface in former times. This principle of uniformitarianism in geological processes had been made explicit by James Hutton in a paper given in 1795, which was brought to general attention by John Playfair early in the nineteenth century. In modern statistical parlance, the same idea goes by the term 'stationarity' of processes. This concept became of crucial, though largely implicit, importance for geographers, in that its adoption permits inference from the present to events in the past, and *vice versa*, and therefore provides a foundation on which to build the analysis of man's relationship to the environment from the evidence of his artifacts.

Darwin's view of the origin of species was also evolutionary, thus reinforcing Lyell's standpoint. In addition, though, Darwin based his ideas on a theory of the relationship of plants and animals to their environment, namely that from one generation to the next only the fittest survive and that a consequential change in genetic structure is slowly effected. Species therefore tend to achieve a perfect adaptation to their environment; if the adaptation is too slow, or the kind of adaptation is disadvantageous, then species will tend to die out.

Lyell and Darwin very quickly became common currency among scientists. Recollect, however, that they were contemporaries of Humboldt and Ritter, the two men responsible for laying the foundations of scientific geographical description. And it was only towards the end of the nineteenth century, and early in the present one, that some of the first attempts at world regionalization were published. Add to these recollections the fact that Humboldt, Ritter and many who followed the early geographical trail adopted an inductive approach to knowledge, and the scene is set for geography's first, albeit somewhat calamitous, attempt at a general theoretical framework to explain patterns of human activity.

Environmental determinism as an idea among geographers is commonly traced back to Ratzel, whose *Anthropogeographie* was published in 1882 and 1889, though in fact the line of paternity goes back at least to Montesquieu's *The Spirit of the Laws*, published in 1748 (see Aron, 1965). As so often, though, it is the disciple rather than the leader who carries things to excess, and determinism is primarily associated with three names – Semple (1911), Huntington (1907 and 1915) and Demolins (1901–3), with Markham's work on climate and the energy of nations coming substantially later (1942). The essence of the determinist stance is that all effects have a cause and that so far as human behaviour is concerned all first causes lie in the physical environment. Actually, Huntington, for example, conceived of two sets of primary causes – environmental and genetic. In a very important sense, this distinction is not material to the present argument, since whether for environmental or for genetic reasons mankind's behaviour was treated as determined by forces outside his control – there was an absence of free will – and the geographer's task consisted of elucidating the part played by the natural environment.

> The only way to explain [civilization] is to ascertain the effect of each of many cooperating factors. Such matters as race, religion, institutions, and the influence of men of genius must be considered on the one hand, and geographical location, topography, soil, climate, and similar physical conditions on the other. This book sets aside the other factors, except incidentally, and confines itself to climate. In that lie both its strength and weakness . . . Many unmentioned phases of the subject have been deliberately omitted to permit fuller emphasis upon the apparent connection between a stimulating climate and high civilization.
>
> (Huntington, 1915, Preface)

Similar ideas had a currency outside the discipline of geography. The Frenchman le Play (Brooke, 1970), commonly regarded as one of the founders of sociology, was by 1856 very active in this field. Using an essentially inductive approach, he visualized societies with common characteristics grouped as:

Savages	Miners and Foundrymen
Herdsmen	Farm workers
Coastal fishermen	Factory workers
Foresters	Distributive workers
	Professional workers

Occupation and the physical environment are clearly closely connected and le Play made a big impact on the development of Demolins' thinking. Jevons, one of the outstanding political economists of his time, developed a strong interest in trade and commercial fluctuations, giving a paper on this theme in 1862. Between 1875 and 1879, he gave a series of papers in which he drew attention to the similarities of periodicities in sun-spot activity and trade fluctuations, offering a possible though rather improbable chain of causal circumstances to link the two sets of phenomena (Jevons, 1909). From his references, it is clear that Jevons' ideas conformed with what had been current for several decades. Beveridge, who went on to become one of the greatest social reformers of modern times, dallied with the same ideas that had attracted Jevons (Beveridge, 1921). Nor can we fail to note Toynbee's (1934–61) monumental study of the rise and fall of civilizations, in which the notion of challenge and response is firmly founded on the challenges presented by the natural environment. Manley (1958) called attention to some 'incontrovertible facts' of climatic change which historians must take into account; and at least one economic historian persists in repeating some of the more naïve determinist views to explain British industrial development in the eighteenth century: 'the temperate climate encouraged people to work hard: it was not so hot that they [the British] became lazy, nor so cold that large-scale industrial development was impossible' (Hill, 1972, p. 18).

Lest we be tempted to dismiss ideas of environmental determinism as quite unfounded, remember the disastrous droughts that have afflicted the countries bordering the southern edge of the Sahara and stretching through Ethiopia to India, from the mid-1960s to the mid-1970s. Evidence has been advanced that this widespread drought represents a long-period climatic change, a trend that is probably 'beyond control by man-induced climate modification. Wholesale movement of population would seem to be indicated' from the sparsely inhabited nomadic territories of Chad, Niger and other states (*The Times*, 28 September 1973). If, as seems probable, the recent droughts are part of a climatic sequence measurable over hundreds of years, this would be consistent with evidence derived from the distribution of ancient dune fields that climatic zones have shifted by hundreds of kilometres during the period of man's occupation of the world (Grove and Warren, 1968; Goudie et al., 1973).

In its unalloyed form, the notion of environmental determinism quickly fell into disrepute. The main reason why this should be so is immediately clear from Figure 6: the causation processes indicated in solid lines were ignored in the determinist model. Two other ideas were quickly introduced. Associated with Febvre (1925) is the idea of 'possibilism', with Nature (still a personified being) providing a

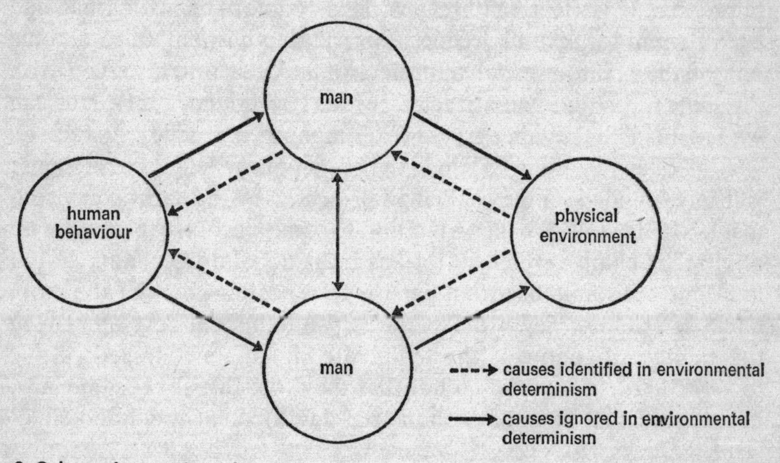

6. Schematic representation of the concept of environmental determinism.

range of possibilities from which man can choose, the possibilities all being equally good. This idea can be interpreted as in Figure 7. Within the inner figure (a circle in this case) all choices are equally good, that is, there is an equal probability of their being selected,

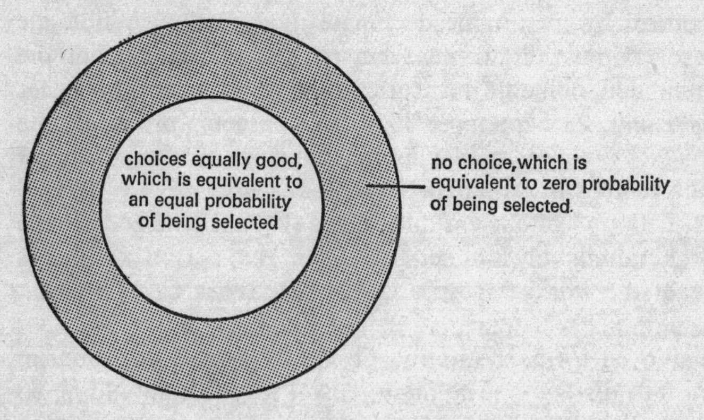

7. Schematic representation of the concept of possibilism.

and outside the demarcated area there is no choice whatsoever, which amounts to a zero probability. Spate (1952) argued for the notion of 'probabilism', that man has choices but that some are more likely to be adopted than others. In this formulation, we may envisage a map of probabilities which is equivalent to how nearly optimal the choice is. Rational men could be expected to select that location for a particular activity, or activity for a particular location, which most nearly approached a probability of 1·0.

The fundamental trouble with this debate was not the issues at stake but the lack of information available and the conceptual climate from which it sprang. This point is clearly made by Morgan and Moss with reference to one part of the discussion:

> If Ratzel were right in maintaining that the soil was the basis of the state, then his political geography had to await another half century of research before his theories could be applied.
>
> (Morgan and Moss, 1967, p. 343)

In conceptual terms, the move from determinism to probabilism represents a retreat from the ideal situation in which determinate solutions are available for all problems to a recognition that the total system is far too complex for determinate answers to be expected in all cases. While it is necessary to accept that in many situations we must think in terms of the probability of events occurring and recognize the impossibility of achieving fully determinate answers, the aim must surely be to reduce the scope for random or stochastic events and increase the area in which high probabilities of correctitude are achieved. One of the major intellectual challenges facing geographers and other social scientists is how to come to terms with the ideas of chance and probability and relate these to explanatory systems.

Stoddart (1966) has made some pertinent points in this context. As we have already noted, Darwin's ideas on evolution had a formative effect on geographical thought, not least as contributing to ideas of environmental determinism. Yet, Darwin's concept of progressive selection of species was applied to an initial situation which was conceived as a chance, or random, pattern, which progressively was given greater order through the evolutionary process. Though as early as 1913 Brunhes considered the ideas of probability in a geographical context, and though important nineteenth-century work in physics, mechanics and other subjects was based on probability concepts, this viewpoint continued to be largely ignored in geographical circles until relatively recently. Perhaps it is to Curry (1964, 1966) that we must attribute the formal and explicit recognition of chance elements in geographical patterns, though he had the wisdom to warn that 'there are many areas of concern to geography for which probabilistic thinking appears irrelevant' (Curry, 1966, p. 41).

Perhaps we can now understand one of the main reasons why the development of statistical techniques has been so important for geography in recent decades. During the period approximately 1920–50, the inadequacies of the determinist school had been recognized. On the other hand, the potential offered by statistical and mathematical techniques that were being actively developed remained largely unnoticed by geographers, many of whom adopted a descriptive, historical approach to unique situations. This frame of mind is represented by the argument advanced in 1954 by McCarty, that the highest level of abstraction attainable in geography would be inductively derived generalizations of spatial patterns and relationships. When geographers did discover the use of statistics, the 'quantitative revolution' sparked enormous excitement as geography advanced in great strides to regain the status as a science that had been in considerable measure squandered in previous decades.

Adopting a statistical, or probabilistic, frame of reference does not of itself solve all our problems; it merely gives us the means with which the better to tackle them. If we refer again to Figure 7, it is clear that considerable problems arise in specifying the *objective* probabilities represented to an individual or group by the natural environment and other human groups, and also in specifying the *subjective* evaluation thereof. Unless the levels of probability, or risk, can be specified independently of the actual behaviour of man, whether measured directly or by inference from his artifacts, there is no means whereby any explanatory models can be made operational. In this respect, it is instructive to examine Pred's 'Behavior and Location' (1967 and 1969) and to ask whether his concept of human actors operating in a two-dimensional space – supply of information and ability to use information – is in fact any advance on the ideas of the determinists. In practical terms, the answer would seem to be 'no', partly for the reason advanced by Morgan and Moss (p. 41) and partly because of the inherent difficulty of measuring accurately people's perception processes (see Brookfield, 1969; Downs, 1970).

Perhaps somewhat paradoxically, it was in physical geography that work was going on which in the long run was to pave the way for further advance in human geography. The geological survey of America produced two outstanding men, J. W. Powell and G. K. Gilbert, who in turn provided the basis for the work of W. M. Davis. Davis' formative publications appeared shortly before and after 1900 (see Davis, 1909) and though his views on the evolution of the landscape were, and still are, disputed (see, for example, Penck, 1925), this is a matter of small moment compared with the mode of analysis that he employed. In 'The Geographical Cycle' (1899) and related essays, Davis attempted to produce a normative view of how the landscape should evolve, hence deriving an ideal against which the real world could be compared. Consequently, he abstracted from reality and produced a much simplified model; while recognizing that ultimately everything depends on everything else, he accepted the need to draw arbitrary though convenient limits to the study in hand. The third, and for our purpose the most important, characteristic of Davis' thinking was that he specified the initial conditions, postulated the processes that would be operational and deduced the developmental sequence that would follow. Consequently, his models provided the means whereby they could be tested, even though he himself did not proceed to make these tests: by detailed examination of actual landscapes to see whether they conform to the predicted morphology; by observation and measurement of the operative processes to see whether they are as postulated.

That this approach was both novel and antipathetic to many senior geographers of the time is briefly but clearly described by Dickinson (1969, pp. 119–22). Hettner in particular reacted very violently against the method 'received from America . . . it seems to me to be wrong in its theory as a whole and in its method in particular' (Hettner, 1921 and 1972, p. xxi). In turn, Hettner was a major influence upon Hartshorne (1939). Nevertheless, in Germany, America and also in Britain physical geographers did accept the Davisian mode of analysis, in particular by setting out to measure actual processes operating in the environment, both by direct observation and by the application of mathematics and physics (though see Chorley, 1970). Five men stand out as particularly important in this context, not only for their own contributions but also for the stimulus they provided to the development of the subject:

Steers (1934)	coastal processes
Lewis (1940 and 1960)	glacial movement and erosion
Bagnold (1941)	wind action and dune formation
Horton (1945) }	stream processes and slope
Strahler (1950) }	development

At least in the British context, the climatological work of P. R. Crowe was also important; one of his pupils was among the first to expand the teaching of statistics to geographers at undergraduate level, at Liverpool in 1957 (Gregory, 1963, vii).

The strength of the tradition thus established in physical geography is attested by two recent works in the geomorphological field, both of which emphasize the need for, and use the results of, the measurement of denudational processes (Pitty, 1971; Carson and Kirkby, 1972). Thus, from about 1940 work in geomorphology began seriously to tackle the problem of 'scientific' explanation by means of rigorous measurement, the testing of hypotheses and the development of theory, so playing a major part in paving the way for the 'quantitative revolution'.

The other major strand of thought that needs to be noted here, though it is discussed more extensively later, may be labelled location theory. Although German workers from the time of von Thünen in 1826 had examined the problems of specifying optimal locations and patterns of land use, it was not until the translation of Weber's book on the economics of industrial location in 1929 and the publication of Christaller's *Central Places in Southern Germany* in 1933 that geographers took a great interest in the deductive and abstract modes of analysis of human phenomena that paralleled the Davisian approach to landforms. In fact, it was not until after the Second

World War that formal location theory was widely known to geographers; and then, with a few honourable exceptions such as Dickinson (1947), only because of the appearance in 1954 of Lösch's *The Economics of Location* and in 1956 of Isard's *Location and Space-Economy.*

Formal location theory thus opened the door for a much more rigorous approach to the solution of problems. It caught the imagination of many geographers precisely because geographers were identifying important issues which could not previously be handled adequately with the theoretical constructs available. By the end of the 1950s, the development of both theory and of statistical techniques made it possible to pose questions in a much more rigorous way than previously was the case.

FROM HARTSHORNE TO HARVEY

The discussion has led us to the period in time when geography began to change very significantly under the impact of quantification. As has been implied in the preceding paragraphs, major intellectual changes occur on a time-scale that must be measured in decades rather than years. Fully to capture the nature of the change that has occurred, it is helpful to pause at this point and compare Hartshorne's *The Nature of Geography* (1939) and Harvey's *Explanation in Geography* (1969). Most seminal works emerge from a body of ongoing work and in a sense are catalysts of thought, a proposition that seems to be true of both these books. Consequently, to have them both open on the desk and to look at them together is probably as good a way as any to capture the essence of what is 'new' in geography.

The Nature of Geography is a difficult book; the argument is tortuous, mainly because it contains implicit assumptions that necessitate a good deal of special pleading. However, for the present purpose it is sufficient to note just a few main characteristics of Hartshorne's work, characteristics that remain in his later publication (Hartshorne, 1960). In the first place, the mode of approach is explicitly inductive, based on a historical review of previous writers, his concern being 'with the establishment and changes in thought concerning the nature of geography, as represented however in geographic works as well as in direct methodological studies' (Hartshorne, 1939, p. 33). Displaying an admirable academic detachment, he continues: 'We will therefore attempt to consider the concepts of previous geographers, so far as is humanly possible, independent of our own particular views' (p. 34). Yet there is no discussion of who shall be classified as a geographer, a classification that manifestly will affect the conclusions which are reached concerning the nature of the subject. Implicit in this approach is the notion that geography – or indeed any other subject – has an objective existence independent of the inquirer. That this was indeed Hartshorne's view may be seen in his quotation – with evident approval – of a phrase from Ritter: 'we must ask the earth itself for its laws' (Hartshorne, 1939, p. 55).

As conceived by Hartshorne, geography is a chorographic science, that is, one that is concerned with the nature of different places, of the way phenomena interact in each place to create areal differentiation. The purpose is to describe and interpret the actual world, and therefore geographers should provide positive analyses of what *is* rather than indulging in discussion of what *will* or *should be* the position; a normative view of an ideal world, or even merely a better one, is essentially alien to Hartshorne's thinking. Above all, therefore, the role of the geographer is to integrate and synthesize knowledge, thus following the German tradition of seeking to identify the whole which has a significance different from the sum of the parts.

With his *Explanation in Geography*, Harvey set himself a goal significantly different from that aimed at by Hartshorne. Instead of trying to identify a subject called geography and its relationships to other subjects, Harvey seeks to elucidate the paths we must follow if we want to add to geographical knowledge in any sense other than the accumulation of facts.

There seemed to me to be nothing wrong with the aims and objectives of traditional geography (indeed they are to be prized and cherished), but as an academic enterprise it had managed somehow or other to hedge itself about with so many inhibiting taboos and restrictions that it could not hope to realize the aims and objectives it had set itself. In particular, geographers were failing, by and large, to take advantage of the fantastic power of the scientific method. And it was the philosophy of the scientific method which was implicit in quantification.

Some people may flinch at the term 'scientific method', so let me make it clear that I interpret this in a very broad sense to mean the setting up and observing of decent intellectual standards for rational argument. Now it is obvious that we can observe these standards without indulging in quantification. Good geographers have always observed them.

(Harvey, 1969, vi and vii)

The text that follows these prefatory remarks has several related characteristics of prime importance to us. In the first place, the distinction, in principle at least, between matters metaphysical and therefore not amenable to proof and those issues for which rigorous methods of inquiry are available is made abundantly clear. Second, and related to this first characteristic, normative statements can be as much a part of geography as positive analysis of how the world, or segments of it, actually works. The third noteworthy characteristic of Harvey's book, which is both implicit and explicit, is that a geographer is faced with problems for which he seeks an explanation. Characteristically, this means defining the problem carefully, identifying the limits of the system that it is relevant to study, specifying the interrelationships of the phenomena and seeking to identify the nature and magnitude of the forces that operate. Finally, and as a consequence, he is primarily interested in probing for general laws about phenomena, to identify patterns and processes that are similar or identical even though occurring in different contexts. To this end, a combination of inductive and deductive reasoning is necessary so that hypotheses may be formulated and tested. By contrast, Hartshorne saw the geographer's goal more in terms of relating unique events to the total system of which they are part and establishing 'generic concepts' by inductive reasoning.

In a nutshell, therefore, Hartshorne and Harvey represent the antithesis between a somewhat mystical or romantic view of geography and a more analytical approach based on accepted canons of scientific method. The former has been surprisingly persuasive in geographical work – as witness the discussion about the nature of regions – and in large measure has its roots in German philosophical traditions of the nineteenth century. The latter derives more from the thinking of logical positivist philosophers and the rationalist views of English-speaking and French scholars. It might fairly be said that the one book represents an end, the other a beginning.

GEOGRAPHY AS A 'SCIENCE'

One of the themes that has clearly emerged as central in the evolution of geography is the problem of finding an acceptable framework within which to offer explanations of phenomena and events. The 'quantitative' revolution has been widely hailed as just the transformation that was required, converting geography into a 'respectable' science (Burton, 1963). Undoubtedly the 1960s have witnessed a major transformation in the subject; today, geographers are probably at least as numerate, if not more so, than most other social scientists. However, the movement to quantification should not be regarded as an end in itself; it is but a means to an end, a symptom of something deeper that is going on within the subject. Harvey (1969, p. 18), citing Chorley and Haggett (1967), remarked: 'The so-called quantitative movement [is] but a symptom of the search of many geographers for some new paradigm.'

The essence of this new paradigm is the model of certain sciences, such as physics, in which the approach to knowledge is by deduction and hypothesis, the testing of hypotheses by carefully controlled experiments in which predicted outcomes are measured against reality, in short by establishing the logical steps of causation and measuring the relative importance of causal factors. In its purest form, this intellectual apparatus yields the proposition that to understand how a system works is equivalent to being able to predict its future states. A classical example is provided by the theories of the motions of the planets and the sun. In the first place, it was for long held that the sun revolved round the earth; secondly, it was supposed that all orbits were circular. Copernicus in 1543 was the first modern scientist to show that a very much simpler and more satisfactory model could be formulated on the assumption that the earth revolves round the sun. About half a century later, Kepler showed that apparent irregularities of motion, especially of Mars, could be accounted for by assuming orbits to be elliptical instead of circular. Thus, we can now make accurate predictions of events such as eclipses, as well as post-dict the positions of planets and stars to check theories concerning the uses of early stone monuments such as Stonehenge. More dramatic, perhaps, is the ability to land men on the moon and get them safely back to earth, a feat that requires extraordinary levels of understanding and ability to predict incredibly complex systems, but which also yielded information on the internal structure of the moon precisely because the orbits of space craft around our satellite differed somewhat from the trajectories that had been predicted.

This idiom of scientific work is in striking contrast to work couched in another form of understanding, which in its extreme manifestation is the empathy that one person has for another. One of the qualities of a good school teacher, doctor or social worker is to relate to the child, patient or client and to understand the emotions and thought processes that are going on. This is a gift of insight that in the last resort is not amenable to formal analysis and verification. In the subject geography, this kind of approach is exemplified by the great regional geographers, who captured the spirit of a place in much the same way as Constable and Turner evoke an instinctive feeling for the scene they painted. Even without this degree of intuition, one can understand how particular events occurred, and why, without necessarily being able to predict their future occurrence. Retrospectively, it is easy to see the significance of routeways from the eastern seaboard of the United States to the interior, or the significance of resources and location for the development of the Ruhr, but how confident could we be in assessing the impact of the third London airport at Maplin, had it been built?

To return, then, to the scientific paradigm, we should note that fashionable scientific method requires of us two things in particular, both derived largely from physics and logic. In the first place, it is axiomatic that, in any model building or theory construction, the reasoning must be internally consistent. In the second place, the outcome of our labours should be the ability to predict the future state of the system under study, given stated assumptions. The success or otherwise of the model that has been constructed is tested by comparing the predicted with the actual outcome. In an ideal world, all work would be directed to both ends simultaneously (that is, consistency and verifiability), but in reality it is often necessary to sacrifice one objective for the other. Furthermore, there is no *a priori* reason to show that all problems will in principle be amenable to the application of both principles, or even of one; nor is it self-evident that the more important problems are the ones to which the 'scientific method' is the most appropriate.

The reader may feel somewhat impatient with this cautious approach to the advantages of modern geography. Lest he be tempted to dismiss these doubts as misplaced, perhaps even harmful, it is worth taking time to look at the agonizing self-appraisal which is currently being undertaken by economists. Economics is far from unique in suffering an attack of grave self-doubt; however, economics is one of the more respected of the social sciences, noted for its logical rigour, the elegance of its theories and the degree of numeracy achieved. Therefore, the experience of this leading discipline must surely be relevant both to those who aspire to emulate it and to those who take a wide interest in intellectual developments generally.

A 1972 issue of the *Economic Journal* carried two articles, both by eminent economists, on the shortcomings of their subject. Phelps Brown wrote under the title 'The Underdevelopment of Economics' and caustically took as his starting point: 'the smallness of the contribution that the most conspicuous developments of economics in the last quarter of a century have made to the solution of the most pressing problems of the times' (Brown, 1972, p. 1). The heart of his indictment lies in the following passage (p. 9):

> In every science the ascending scale of intellectual status tends to be one of rarefication: the more abstract, the more rigorous, the more general, so much the more distinguished. This is natural, because distinction is conferred by rarity, and few of us are capable of soaring into the empyrean of abstraction, whereas there is a saying that no man is so short that his feet do not reach the ground.

Worswick, in the same issue of the *Economic Journal*, poses the question: 'Is progress in economic science possible?' and, though he concluded that the answer is in the affirmative, was careful to add that such progress would be slow. His views can be readily summarized:

> The more the impression is allowed to persist that economics is an exact science, or if not already one, then with the aid of mathematical models and the computer is about to become one, the more damage will be done to the subject when it fails to live up to exaggerated expectations.
> (Worswick, 1972, p. 84)

Another eminent economist (Champernowne, 1973, p. 908) has even referred to economics as a 'pseudo-science'.

A substantial part of the reason for these difficulties and doubts is contained in a joke retailed by another economist (Boulding, 1970, p. 101):

> There is a story that has been going around about a physicist, a chemist, and an economist who were stranded on a desert island with no implements and a can of food. The physicist and the chemist each devised an ingenious mechanism for getting the can open; the economist merely said, 'Assume we have a can opener'.

The reader will readily perceive the nature of the dangers that lie ahead of the 'modern' geography unless its practitioners take heed. The first of these dangers is that to achieve intellectual rigour it is necessary to assume away the inconvenient complications, thereby developing theories that can have little or no conceivable practical application, though elegantly and logically constructed. The field of economics is littered with non-operational theories of economic growth and development; in the geographer's domain one may note the inadequacies of location theory (Chisholm, 1971a), and the unresolved dispute as to whether urban hierarchies should be based on a $K = 3$, $K = 4$ or $K = 7$ system (Christaller, 1933 and 1966; Garner, 1967; Berry, 1967). The opposite danger is to concentrate on those problems which are amenable to satisfactory scientific analysis but which in many cases may be of trivial importance.

What, then, should our view of geography as a science be? In the first place, it seems essential to establish that there is not one single methodological standpoint for which it can be claimed that there is universal validity, whereby anyone who adopts another viewpoint is regarded as universally wrong. Once that is said, however, it is also abundantly clear that where possible there should be accepted procedures of reasoning, methods of investigation and so on, so that dispassionate workers may replicate the efforts of others and check their conclusions. Thus, the model of scientific investigation which requires internal consistency of reasoning, the verifiability of conclusions and the ability to predict is clearly the preferred mode of analysis at which we should all aim. While it would clearly be wrong to insist that only those problems which can be framed in these terms should be investigated, it is equally clear that we should strive to expand the area of human experience to which the 'scientific' paradigm applies. In terms of Figure 8, the ideal position is the bottom left-hand corner of the cube, which represents a three-dimensional space. Much of the work in the natural sciences, engineering, physics and chemistry lies very near this ideal point (marked by X). One of the troubles with economics is that a high proportion of work (theories of growth and inflation for example) lies near the point marked Y.

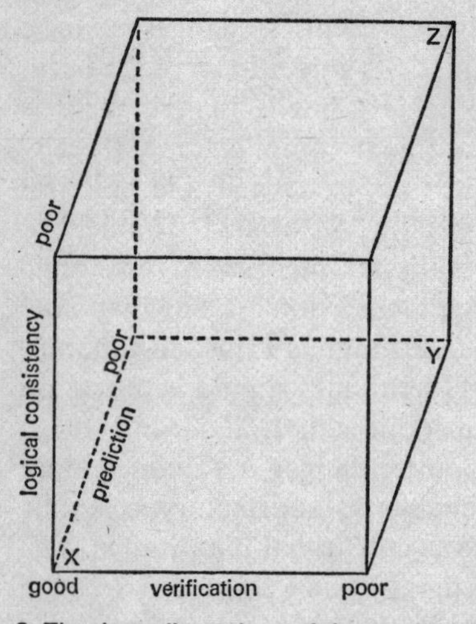

8. The three dimensions of theory. (For the significance of X, Y, and Z, see the text.)

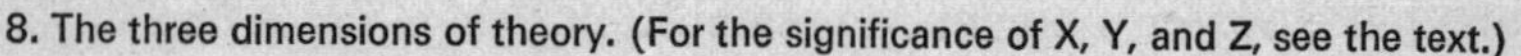

The major change in geography that has occurred since the last world war is associated with quantification and is the acceptance of the ideal for scientific analysis. Implicit in this acceptance is the desire to expand the section of the cube in Figure 8 that is labelled 'good' and diminish the part identified as 'poor'. But in so doing, we must not make the mistake that many made in the nineteenth century of assuming that if we know enough then everything will be explained. There are several reasons for making this cautionary proposition, but three in particular deserve attention since each on its own is a sufficient condition:

1. A rationalist is one who 'regards reason, rather than sense, as the foundation of certainty in knowledge' (*Shorter Oxford English Dictionary*). To be truly rational, however, it must be accepted that there is no *a priori* proof that everything is amenable to reason.

2. The sheer complexity of phenomena may defy analysis owing to the vast quantity of information that must be collected and processed. The American Apollo series of manned flights to the moon required a computer system at ground control that exceeded the total computer capacity then installed in the whole of the United Kingdom. Since resources are scarce, it is evident that many problems must remain unsolved for want of the analytical capacity.

3. Heisenberg's principle of indeterminacy applies not only to minute particles of matter but also much more generally. In examining rock samples from the moon for evidence concerning the moon's origin, there is an element of doubt whether the materials have been contaminated with earth-derived constituents. Similarly, social survey polls, as at elections, may well influence the event they are designed to monitor.

It is therefore of some interest to inquire whether there is not another paradigm to which we can turn. In asking this question it is not intended to suggest that the model of scientific inquiry which we have been discussing should be rejected. Rather, as one moves toward the upper-right part of Figure 8 – indicated by Z – the scientific paradigm is manifestly inadequate and yet it would clearly be desirable to have some framework within which to work. An answer given by Mr Andrew Shonfield, as Chairman of the Social Science

Research Council at a conference held at Nottingham in 1971, has been endorsed by Ward (1972). Shonfield had in mind that for the social sciences generally it may be the case that 'scientific proof' is impossible to attain but that informed opinions can be given. The analogy is with the way that an individual or a corporate body can take legal advice on a point of law. Where there is a good case-history relevant to the issue at hand, and the law is unambiguous, then the advice tendered may be regarded as representing a certainty. In other cases, where these conditions are not met, the opinion of one counsel may differ from that of another and the only method of resolution lies in taking the matter to court. Furthermore, the judgement and sentence handed down by a lower court can be amended or reversed by a higher court. The essential point is that the legal profession operates certain recognized procedures and employs established modes of reasoning to relate a particular case to a general corpus of law and other cases which have already been tried. Thus, there is a highly developed process for obtaining opinions and judgements.

Ward elaborates on this theme by noting that in the judicial machinery there is a fairly explicit incorporation of value judgements. Senior judges do change, or at least modify, the application of the laws by incorporating their own moral/metaphysical viewpoints; Lord Denning in Britain and Chief Justice Warren in the United States are two modern examples, while Jeffreys, who held the seventeenth-century Bloody Assizes, is notorious in English history. Perhaps more interesting is Ward's argument that the 'validation' of law consists of the replication of similar cases treated by the same judicial processes. If each case is regarded as equivalent to a scientific experiment, then the accumulation of results builds up to 'proof' that something is either right or wrong. The nature of this 'proof' differs in one important respect from that derived in formal scientific experiments, in that one case in law is not independent of those cases which have preceded it. With this reservation, it seems reasonable to postulate a fairly close analogy between the two processes for eliciting the true situation, though they apply in different domains.

All this may seem a far cry from geography. To summarize the issues examined above, let us take a commonplace proposition. In the use of regression analysis, it is widely accepted as dangerous to extrapolate the regression equation beyond the maximum and minimum values used in fitting the equation, that is, beyond the domain of the data. In the same way, it is asking for trouble if a particular paradigm of inquiry is used in situations for which it is not appropriate. Alternatively, if a single paradigm is employed, then we must be explicitly aware of the fact that we are shutting our eyes to a very large array of problems. Most of the remainder of this book is concerned with an examination of recent developments in geography, which have centred around the discovery of quantification and the realization of the potential of the 'scientific' paradigm, the hypothetico-deductive frame of reference. Especially in the concluding chapter, however, the question is discussed whether this 'scientific' approach really is adequate. Some geographers have recently gone on record to argue that it is not. However, before we reach this question, there is a good deal of ground to be covered, to which attention is given in the following chapters.

3. Static Patterns

As Bunge (1966, p. 2) observes, 'it is useful to divide science into three elements: logic, observable fact, and theory'. An adaptation of his schematic presentation of the relationships between these three elements is shown in Figure 9. Facts may be arranged in various

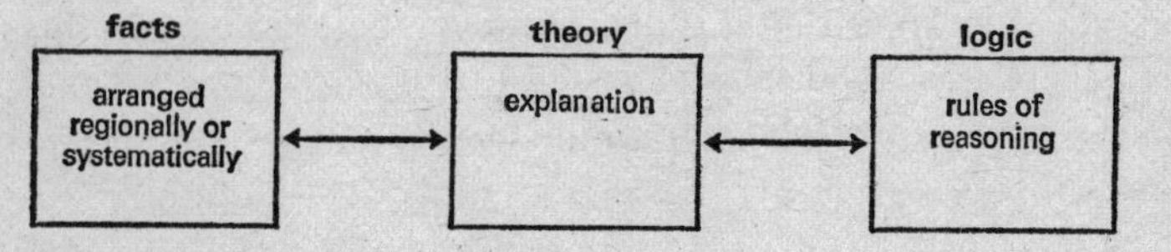

9. The relationships of fact, theory and logic.

ways, as in Berry's geographical data matrix, but what items are recognized as facts is both conditioned by and in turn conditions our theoretical, or explanatory, constructs. These in turn have a reciprocal relationship with the rules of reasoning, as between the boxes labelled 'theory' and 'logic'. While it is often convenient for description (facts) to precede explanation (theory), it must be recognized that the description is itself, whether explicitly or implicitly, conditioned by some hypothesis or theory. However, the key proposition relevant in the present context is that the range of *possible* theories is limited by the means available to us for describing the patterns presented by the 'facts' that exist. Thus, a quite fundamental part is played by description and it is essential to review some of the main modes employed, other than verbal techniques.

Accurate description of geographical patterns involves two closely related issues, both being special cases of the need to transform data from one form into another. The first is a problem that we have already encountered, namely the need to choose a metric of space, usually conceived as some kind of co-ordinate system whereby locations can be mapped relative to each other. In the second place, data rarely come to hand ready-made in the metric of space with which we wish to work and therefore they must be manipulated – transformed – in a suitable manner. Both issues are amplified in the succeeding paragraphs.

We have already seen in Chapter 2 that there are immense problems in transforming the three-dimensional spheroid of the earth into a two-dimensional map. Let us assume that for the purpose in hand this problem has been solved. The most common procedure then is to use a system of reference grids which for small areas take the form of cartesian co-ordinates, regularly spaced reference lines that are perpendicular to each other and allow locations to be defined relative to some arbitrarily selected origin. The British Ordnance Survey uses such a system, known as the National Grid, with the origin lying south-west of the Isles of Scilly. However, there is no need to limit consideration of space to the metrics which are based on physical distance or the geometry of Euclid; it is equally valid to

> conceive of events located in other kinds of space for which the units of measurement may be time, cost, some index of convenience, or some other basic concept (Watson, 1955). An important early example of the mapping of accessibility in terms of time is the evidence submitted by the Royal Geographical Society to the Barlow Royal Commission on the location of the industrial population in Britain (Taylor, 1938). Another of the pioneering uses of this approach still remains a classic: Harris (1954) examined access to markets as a factor in the location of manufacturing industry in the United States and conceived of space in terms of costs of movement, whereby the relative cost of access to the whole market can be assessed for each location to find that which will minimize marketing costs.
>
> (Chisholm, 1971b, p. 12)

Both Taylor and Harris drew their isopleths on a conventional map of the respective countries; but it is easy to conceive of maps re-drawn so that locations are related to journey time (or cost) from a specified origin. Bromhead (1973, p. 93) did just this, presenting a map of Europe in terms of expected rail travel times from London in 1980–85, as part of his attack on the decision (subsequently revoked) to build a third 'London' airport at Maplin, arguing that many air passengers on London–Europe routes will find rail travel competitive in overall journey time.

An interesting further development of this approach is to draw maps in which the areal units are distorted so that their area is proportional to some phenomenon other than their acreage. The technique can be attributed to Raisz (1938) who published cartograms in which the areal units were represented as rectangles, the areas of which were proportional to total population, value added by manufacturing, etc. Development of the technique has been mainly by American geographers but has been most interestingly applied by Hollingsworth (1965) to map the electoral districts of England and Wales, the area of each constituency being proportional to the number of voters. The idea has also been taken up for the presentation of medical statistics (Forster, 1966) and has been more generally discussed by Häro (1968).

(Chisholm, 1971b, pp. 12–13)

Transforming space from the familiar 'geographical' form of a co-ordinate system linked to longitude and latitude, in which locations remain unchanging over time, introduces considerable variation in relative locations with the passage of time. As Janelle (1968) has shown, the journey between London and Edinburgh now takes only one hundredth of the time required in 1658, or a little less than one tenth of that needed in 1800; in the period since 1800, the convergence of Boston and New York has been taking place at about the same rate as the convergence of London and Edinburgh. In both cases, of course, the distance in kilometres has not varied significantly.

The general nature of the space transformation problem is very neatly demonstrated by Bunge (1966, p. 53). Figure 10 depicts four ways in which a system of places and routes can be portrayed. Figure 10(a) represents the actual locations and the connecting routes, whereas in 10(b) the routes have been converted into straight lines. In 10(c), the relative locations have been changed and the shape of the connecting routes altered; however, careful inspection will show that 10(c) gives an accurate picture of the way in which the places are connected in terms of graph-theory language of edges and nodes (see p. 95). In 10(d), the information has been transformed into matrix format; each occupied cell represents one of the links (edges) that connects a pair of nodes. This matrix conveys the minimal

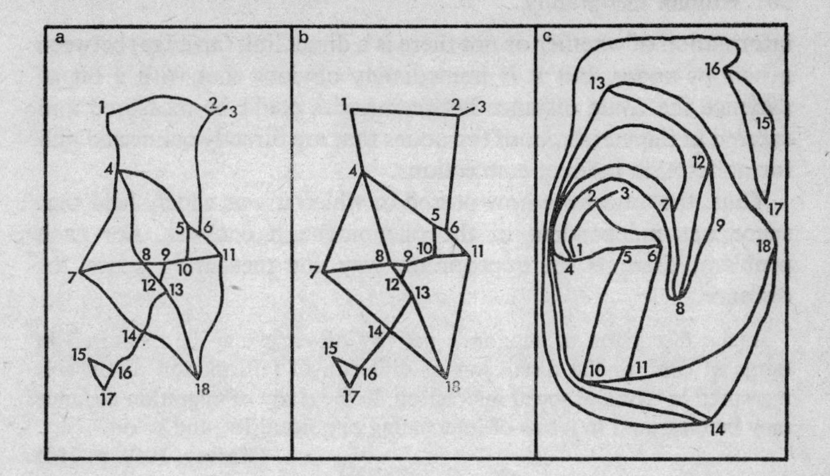

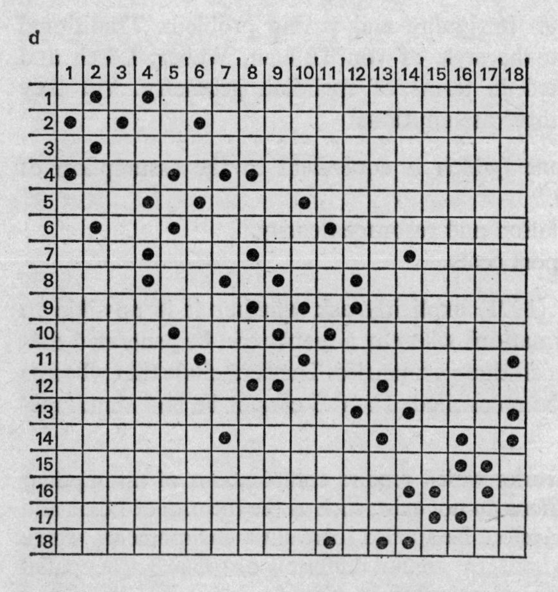

	1	2	3	4	5	6	7	8	9	10	11	12	13	14	15	16	17	18
1		●		●														
2	●		●			●												
3		●																
4	●				●		●	●										
5				●		●				●								
6		●			●						●							
7				●				●						●				
8				●			●		●			●						
9								●		●		●						
10					●				●		●							
11						●				●								●
12								●	●				●					
13												●		●				●
14							●						●			●		●
15																●	●	
16														●	●		●	
17															●	●		
18											●		●	●				

10. Transformations: networks and matrices. (*Source: Bunge, 1966, p. 53*)

information of whether or not there is a direct link (an edge) between a pair of nodes, but it is immediately obvious that with a bit of patience the route distance between nodes could be measured and entered in the matrix, both for nodes that are directly connected and for nodes with indirect connections.

Thus, the days have now passed in which it was widely held that space was independent of the phenomena it encloses. For each problem, there is an appropriate way for measuring space, or distance.

> In the discussion of economic activity distance may be measured in terms of cost, in the discussion of diffusion of information distance is measured in terms of social interaction, in the study of migration distance may be measured in terms of intervening opportunities, and so on.
>
> (Harvey, 1969, p. 210)

Here, then, arises an intriguing and vexing problem. Traditional location theories, in the style of von Thünen, Weber, Lösch and Isard, are formulated in terms of euclidian geometry. Yet they commonly contain three assumptions:

1. A uniform plane (which is equivalent to the assumption of euclidian geometry).
2. Uniform population and resource density.
3. Uniform transport costs.

Angel and Hyman (1971) explicitly ask whether it is possible to conceive of transformations whereby a metric can be generated that simultaneously gives all three properties. They conclude that whereas 1 and 2 above can be reconciled, 1 and 3 cannot be had simultaneously:

> The geographic theories which require combinations of assumptions for which transformations do not exist, such as the theories of Lösch and Christaller, cannot therefore be applied to realistic environments, where densities and costs of transport are not uniformly distributed. The spatial assumptions of these theories need to be relaxed.
>
> (Angel and Hyman, 1971, p. 32)

This problem is directly paralleled by the impossibility of constructing a map for any significantly large portion of the globe that simultaneously retains more than two of the four properties – correct scale, area, shape and azimuth.

The reader will perceive that there are two related problems concerning the way in which space is to be used in geographical studies:

1. The concept of space to be employed in a particular case.
2. The symbolic representation of that space.

A major focus of geographical research, both in the past and to the present day, has been to explore these problems. However, before we examine some of the substantive work designed to elucidate spatial patterns, there is one further point to have clearly in mind. Harvey (1968) has noted that the processes which operate to create spatial patterns function at different scales. For example, with an estimated 5 million spiders per hectare in summertime Britain (*Observer*, 19 August 1973), an analysis of their spatial organization would require the researcher to distinguish locations to an accuracy of at least one metre and probably in terms of centimetres. On the other hand, a resolution at one kilometre will usually be adequate for studies of the spatial distribution of man's settlements. Therefore, just as an appropriate level of resolution should be adopted in mapping out a particular pattern, so may the pattern that is discovered depend on the spatial scale employed.

To summarize this discussion, it will be helpful to make clear the notion that a 'map' is to be conceived not in the colloquial sense of the maps which one can buy in bookshops or find in atlases, but as a logically based system for showing the relationships of phenomena in space, whether the space be euclidian or derived from some other geometry. The form of the map implies one or both of two things:

1. Some theory about the spatial ordering of the phenomena.
2. Some purpose to which the map will be put, for example, for the inductive development of theory.

Bunge (1966, pp. 33–5) confuses the issue when he suggests that a map is a theory; it is merely the logical basis on which either to represent or build theory. And so we have the paradoxical situation noted by Dacey (1973, p. 131):

The basic methodological statements by Hartshorne (1939), Bunge (1966) and Harvey (1969) almost totally ignore the study of map distributions and its role in the construction and testing of geographic theory.

SINGLE VARIABLE PATTERN

The recognition of a pattern – whether spatial or otherwise – depends on two fundamental processes. The first is the isolation of the phenomenon in which one is interested from all others, which is essentially the major step in getting rid of what is now commonly called 'noise', a term denoting unwanted information. The second step is the identification of the major regularities in the data set, which is equivalent to generalizing the complex pattern into its basic components – a smoothing or filtering process which seeks to eliminate the remaining 'noise' in the data. Early attempts to portray spatial patterns were generally aimed at the improvement of description rather than the testing of theories and relied very much on subjective evaluation; during the past few decades, techniques with a stronger objective basis geared to the development of theory have been devised and have become popular.

Only with the publication in 1938 of Raisz's book *General Cartography* was the disparate material on techniques for recognizing spatial patterns drawn together, at least in the English language. While his book represents an admirable survey of the then existing state of the art, it is noteworthy that most of the techniques he discussed remain in vogue today (Monkhouse and Wilkinson, 1952; Dickinson, 1963). The major methods for presenting spatial patterns can be classified as follows:

1. Isopleth (line of equal value)	contour lines, rainfall, temperature, etc.	equivalent to a continuous surface
2. Choropleth (area of uniform value)	population density, land-use proportions, land prices, etc.	ratios of two variables, a discontinuous surface

3. Relief presentation (including contours)

4. Miscellaneous – dot maps, pie graphs, space transformations, etc.

Major interest in the present context attaches to the first and second categories. Both isopleth and choropleth techniques are highly satisfactory ways of representing spatial patterns, and which one is used depends mainly on the nature of the data available. For example, rainfall records are available for specific places (recording stations) and the most direct way to portray the precipitation pattern is to interpolate isohyets, or lines connecting places with equal precipitation. On the other hand, population statistics refer to a defined area, allowing density ratios to be calculated and mapped. However, whichever technique is adopted it is important to be able to compare one pattern with another, since it is only in this way that explanations can be offered for the distributions that one is specially interested in. Until relatively recently, such comparison had to be done in a subjective manner, except in the relatively rare cases in which the relevant data were available for identical recording stations or statistical areas, when standard regression techniques can be used (McCarty et al., 1956). Clearly, this was an unsatisfactory situation, but fortunately recent decades have witnessed sustained efforts to provide tools for the making of formal comparisons, so that the research worker may set up properly formulated hypotheses and test them in a rigorous manner, even though the original data are not available for identical points or areas.

The basic issue is how to transform one map into a form that will permit proper comparison with another. Three kinds of comparison are involved:

isopleth/isopleth
choropleth/choropleth
isopleth/choropleth

Two of the first geographers to recognize the problem and also offer a solution were Robinson and Bryson (1957). To illustrate their ideas, they asked the question: to what extent is variation in population density in Nebraska correlated with spatial differences in annual precipitation? From the population and precipitation data available, they constructed two isopleth maps, which can be viewed as representing two continuous surfaces. These they superimposed over a regular grid. For each of the points identified by intersections of the reference lines (as in Figure 11(b)), numbering twenty-six in all, they estimated the population density and amount of rainfall. They did this by interpolation between the isopleths. Equipped with the twenty-six pairs of observations, it was then a straightforward matter to calculate the degree of association between the variables by means of standard correlation techniques. The key step in this procedure is to produce continuous surfaces from which data are extracted for specified

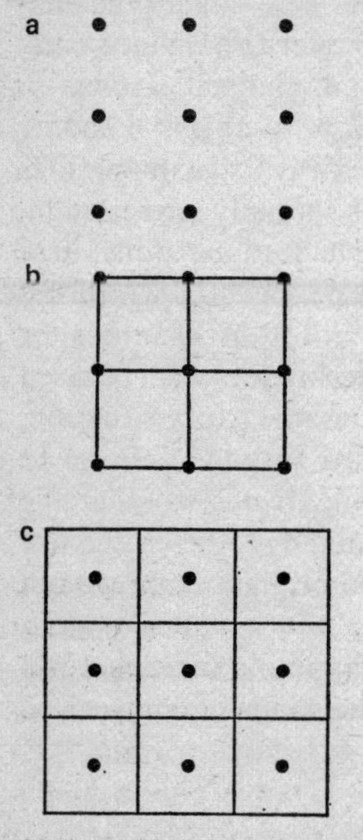

11. Point patterns and lattices.

sample points. The formal comparison is then performed by using the values so estimated for the specified points.

Somewhat earlier than the paper by Robinson and Bryson, Hägerstrand in 1953 published in Swedish his *Innovation Diffusion as a Spatial Process* (Hägerstrand, 1953 and 1967), a work of seminal importance in which considerable use was made of a regular lattice to partition the study area. At about the same time, plant ecologists were working on nearest-neighbour measures for the analysis of plant communities (King, 1969a, p. 89). Also, modern computers were beginning to come into use at the end of the 1950s, and computers operate with data in matrix format. Thus was the scene set for the general acceptance and development of the following proposition. Figure 11(a) shows nine locations set out in a regular array. These locations can be defined as the points of intersection of regularly spaced reference lines or cartesian co-ordinates – as in Figure 11(b) – or alternatively may be assigned to the squares defined in Figure 11(c). Therefore, given data suitable for either isopleth or choropleth mapping, the information can in principle be transformed into a regular point pattern or matrix form. Of course, where the original data relate to an irregularly spaced set of points or areas, there are considerable computational problems, both at the theoretical and practical levels, which have only recently been solved. Thus, one of the major research efforts in geography during the 1960s and 1970s has been the exploration of the opportunities offered by point-pattern and lattice techniques of analysis.

LATTICES

An extended discussion of the properties of lattices and the uses to which they may be put would be out of place in the present context (see Dacey, 1965; Cliff, undated; Bassett, 1972). On the other hand, some assessment of their utility is important given that they have been on the geographical scene for only a relatively short time. The simplest form of lattice consists of regular squares, but in fact a lattice may consist of regular shapes of a wide variety, triangles and hexagons for example, and also irregular shapes. For the present purpose, we will confine attention to the regular square lattice. Three main uses of lattices may be identified for the purpose of exposition, though in varying degree these uses are interrelated.

First among the three uses of lattices is the one that we have already considered briefly, namely, measurement of the spatial association of two or more phenomena. With data in matrix format, the whole suite of correlation techniques can be brought into use, the particular method (least squares or rank order correlation, chi-squared, etc.) depending both on the data and the purpose of the study. The second use of lattices follows from the first and is especially interesting, since it opens the way to solving a very teasing problem of geographical analysis. Product-moment correlation analysis is based on the assumption that the regression residuals are independent of each other. Although there are well-established techniques for testing serial (that is, temporal) correlation of residuals, it is only recently (Cliff and Ord, 1969) that a method has been developed to test for spatial autocorrelation. Once the regression residuals have been obtained, each cell in the lattice can be assigned to one of two classes, positive and negative (called black and white respectively). If the residuals are not spatially autocorrelated, then the black and white cells will be randomly distributed within the lattice. On the other hand, if there is spatial autocorrelation this will be manifest as groupings of black and white cells. The formal test depends on the idea of contiguity, conceived as the joins between adjacent cells, these being Black-Black, White-White and Black-White. For a given number of black and white cells, it is possible to calculate the number of BB, WW and BW joins to be expected if the cells are randomly distributed and compare this expected distribution with that which is actually observed.

This idea has been extended for the formal comparison of two distributions, where the purpose is to test how well some theoretically derived distribution fits the real-world situation. Although the particular context of this work was to compare simulated diffusion patterns with the actual pattern of diffusion of specified phenomena, the idea has a general utility.

> First it is necessary to construct a map of the differences between the theoretical and actual patterns. One way of preparing such a map is to create a three-colour map in which each county [lattice square] is coloured, say, black, B, if the theoretical and actual values are 'the same' in the county; white, W, if the theoretical value is 'greater than' the actual value, and red, R, if the theoretical value is 'less than' the actual value. Interest is then centred upon the distribution of the W and R counties in the differences map. If the theoretical map corresponds well to the actual map, the W and R counties should be randomly distributed in the map of differences.
>
> (Cliff, 1970, pp. 143–4)

In the text that follows, Cliff derives a test statistic embodying the idea which he states so clearly.

Lattices are fundamental to any work in which some spatial process is postulated which is to be modelled, the spatial patterns so derived being compared with the real-world situation. This approach is now commonly adopted when the researcher is not able to observe the process directly but can make reasonable inferences concerning its nature. If he then constructs an operational model incorporating his assumptions about the process in question, comparison with the relevant real-world patterns will enable him to conclude whether or not the postulated processes are plausible. The major application of this approach has been in the field of diffusion studies (see p. 162), prompted in large measure by Hägerstrand's work.

The third major application of quadrat analysis should perhaps have been discussed first, since it logically precedes the other two uses; on the other hand, reserved to the last, it makes a convenient bridge to the section which follows. A simple example is provided by the distribution of houses within a study area: are they randomly distributed (dispersed) or are they clustered in hamlets and villages? If each cell in the lattice is called black or white according as dwellings are present or absent, contiguity measures will allow one to describe the pattern as clustered or random. Such a description is clearly useful for making comparisons between areas, as well as the situation in the same area at two or more points in time. It may also be helpful in probing the nature of the processes which have generated the observed pattern.

POINT PATTERNS

Nearest-neighbour techniques of analysis date very largely from ecological work published in 1954 and taken up by both M. F. Dacey and L. King early in the next decade. Instead of viewing space as divisible into finite areas comprising a lattice, geographical space is conceived as a set of points which for purposes of analysis are supposed to be infinitely small. Such a view of space is clearly reasonable for studies of settlements and their distribution – the classic area for geographical applications of nearest-neighbour techniques – as well as for studies of industrial localization, and the distribution of plants and plant communities: it allows one to determine whether and to what extent a given distribution departs from randomness in the direction of either clustering or regular dispersion (see Figure 12).

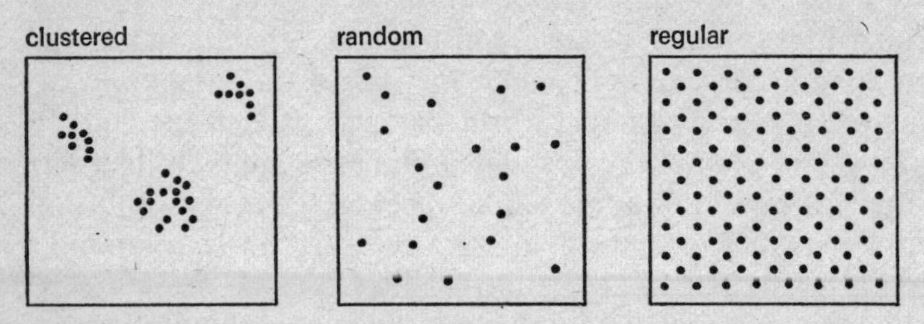

12. Regular, random and clustered point patterns. (*Source: Dacey, 1973, p. 132*)

King (1969a) and Morrill (1971) present very useful summaries of the nearest-neighbour technique, the principle underlying which is quite simple. With a regular distribution of points, as in Figure 11 and the right-hand part of Figure 12, it is clear that for any point the distance to its nearest neighbour is identical to the nearest-neighbour distance for all other points. If, then, the size of area is known and the number of points to be regularly spaced is also known, then it is manifestly a fairly straightforward task to calculate what the nearest-neighbour distance will be. At the other extreme, a clustered pattern will be characterized by nearest-neighbour distances that approach zero. In the third case, of a random pattern, there will be a variety of nearest-neighbour distances, the distribution of which conforms to a special rule of probabilities, known as the Poisson distribution. Thus, with three characteristic 'norms', point patterns can be economically described.

This descriptive technique has a clear family relationship to quadrat analysis, as previously discussed. Perhaps the major significance of both approaches to the identification of spatial patterns is the explicit recognition of the random, or chance, element in geographical distributions and the fact that the technique provides a way for probing the significance of randomness. However, as Harvey

(1968) and Dacey (1973) have noted, both nearest-neighbour and quadrat analysis suffer from serious limitations, not the least being the problem of scale. In Figure 13,

each of the quarters contains one major town and twelve lesser ones. Taking the four squares together, it is immediately evident that the towns are grouped along a NE to SW axis and are not spread uniformly over the whole territory. At this scale, there is clearly a substantial element of clustering. However, examination of the SW quadrant suggests a random scattering of the towns, whereas in the NE the settlements are arranged in a regular lattice.

(Chisholm, 1971b, p. 47)

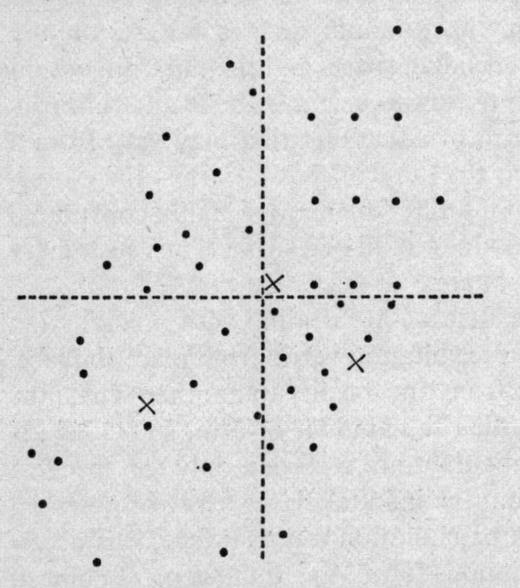

13. How pattern description varies with scale. (*Source: Chisholm, 1971b, p. 48*)

Clearly, the description of the pattern changes as the scale changes. (See also Haggett, 1965, pp. 263–76; Chisholm, 1966, p. 37; Harvey, 1968.)

SMOOTHING SPATIAL SERIES

Even when various transformations of data have been performed – isopleth and choropleth maps, quadrat or nearest-neighbour analyses, etc. – many phenomena present confusing spatial patterns that are difficult to identify visually. One of the techniques that has been developed in recent decades to make spatial patterns more comprehensible is to fit a trend surface to the data. Trend-surface fitting may be conceived as a three-dimensional form of regression equation, the 'surface' being the plane of best fit. Two of the dimensions are orthogonal axes representing north–south and east–west, or any other axes at right angles in euclidian space, and the third dimension is height – or some surrogate notion of height – above an arbitrarily defined datum level. The form of equation fitted may vary from a simple linear version through the range of power terms, the corresponding surfaces ranging from a tilted plane to a highly convoluted surface. In considering the goodness of fit of a surface, it is customary to use the criterion of least squares.

Imagine, for the moment, that we are dealing with precipitation data. The annual amount of precipitation varies from place to place, with a continuous gradation from one locality to another. Thus, the real-world 'surface' representing the amount of rain, snow, etc., is continuous. However, climatic data are recorded only for specified points, as samples drawn from the infinitely large array of potential records. Thus, to compute a precipitation trend surface, it is necessary to use data in a discontinuous form. As a quite general point of procedure, therefore, it is common to assume that data are discontinuous in order to calculate a continuous surface that will generalize the real-world continuous surface. The more closely spaced are the points, the less is the loss of accuracy but the greater the amount of computation involved.

Trend-surface mapping was first developed by geologists in the mid-1950s (Krumbein, 1956; Miller, 1956; Grant, 1957). Haggett and Chorley were perhaps the first geographers to appreciate the possibilities of the technique and since the appearance of their deservedly noted papers (Haggett, 1964; Chorley and Haggett, 1965) geographers have found trend-surface mapping of considerable value as part of their kit of statistical tools. For example, in their study of the changing pattern of urbanization in England and Wales, Hall and his colleagues (1973, pp. 172–3 of vol. I) fitted trend surfaces to the percentage *change* in total population for the periods 1951–61 and 1961–6. The resulting maps show an unmistakable shift of the locus of highest growth from near London to Kent – that is, south-eastwards – while the north-westerly regions of declining population shrank in the opposite direction. At a glance, therefore, one can see that there has been a 'flattening' of the surface, or an equalization in growth rates, between the two periods.

However, as Norcliffe (1969) points out, it is only too easy to violate some of the underlying assumptions of trend-surface mapping, notably in the context of spatial autocorrelation of the data (see p. 66). Furthermore, because trend surfaces are fitted by the regression technique of least squares, there will be residuals both above and below the fitted surface. In the geomorphological context, where the intention is usually to identify an erosion surface, and hence residuals *above* the surface are not acceptable (except in the special case of monadnocks), it seems more appropriate to think of an envelope surface (Tarrant, 1970). This would just touch the isolated areas of high land identified as remnants of the erosion surface.

In the field of human geography, one of the more important competitors to the trend-surface technique was developed by Tobler (1967). With time-series data, it is common to smooth fluctuations by employing a running mean, of three, five or some other odd number of time periods, the average value being assigned to the central year. Tobler's notion is based on partitioning space into a regular lattice, for each cell of which a single value can be estimated for the phenomenon in question. In the simplest case, the smoothing operation consists of assigning to each cell an average value derived as the mean value for that cell plus the eight adjacent cells. Thereby, local extreme values are smoothed out and the general configuration of the pattern can be more clearly seen. The nature of this filtering process can be seen in Figure 14 and compared with the trend-surface approach depicted in Figure 15. The great advantage of Tobler's approach over fitting a trend surface is that the former

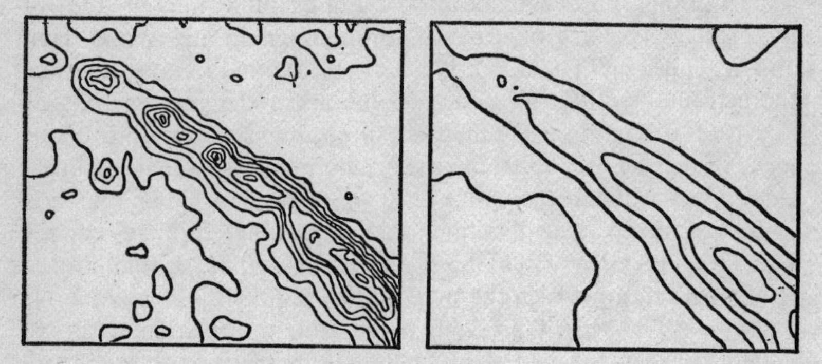

14. Filtering spatial patterns: original pattern on the left, simplified by spatial averaging on the right. (*Source: Tobler, 1967, p. 276*)

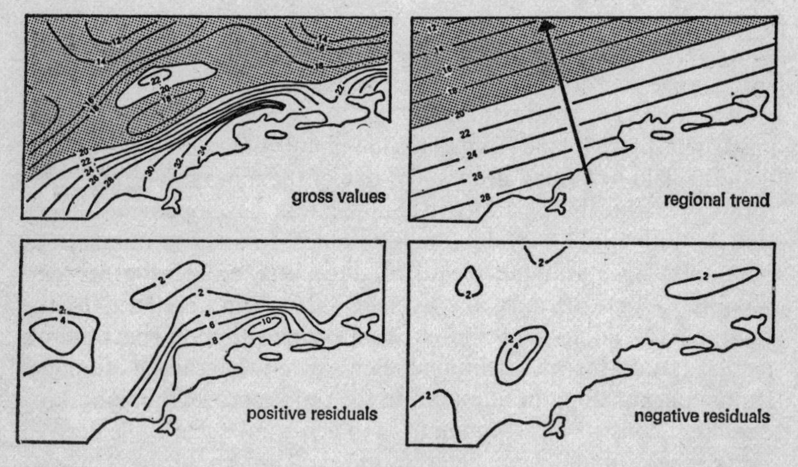

15. Trend-surface mapping of forest distributions in part of Brazil. (*Source: Haggett, 1964, p. 372*)

implies no explanatory theory underlying the observed spatial distributions, whereas the latter does (Norcliffe, 1969).

A final comment on the filtering and smoothing of spatial data is relevant in the context of Chapter 5. While it is a necessary step in analysis to describe observed patterns, the ultimate goal is the derivation of explanatory theories. The question then arises concerning the extent to which one spatially located observation is independent of all other spatially located observations. Consider for a moment events occurring in time, such as the growth of a country's population. It is quite clear that the population in 1974 is very closely conditioned by the total in 1973, and only a little less closely by the number in 1972. There is a strong temporal autocorrelation involving several tens of years, and a difficult problem arises in deciding just how many time periods to include in the analysis. On the other hand, the direction of causation is quite clear: population in 1973 affects that of 1974, and not *vice versa.* In the spatial context the geographical extent of autocorrelation poses problems analogous to those in the temporal domain; however, in the spatial case the difficulties are compounded by uncertainty regarding the direction of causation, or indeed whether there is mutual interdependence. Thus:

> It is clear that many of the methods, which prove to be useful in dealing with time series, are vastly more complicated when generalized in an attempt to deal with data from spatial processes.
>
> (Granger, 1969, p. 13)

One of the great intellectual problems currently facing geographers and other spatial analysts is the extent to which observed spatial regularities are the result of spatial autocorrelation. It was only in the late 1960s that a real beginning was made in attempts to cope with this problem (Cliff and Ord, 1969; Granger, 1969; Bassett and Norcliffe, 1969).

REGIONS AND REGIONALIZATION

> Regions occupy a central position in geography and most of the 'classics' of geographical literature . . . are regional monographs. Athough regions

have come under some heavy crossfire . . . they continue to be one of the most logical and satisfactory ways of organizing geographical information.
(Haggett, 1965, p. 241)

In terms of Figure 16, we may fairly describe the history of regionalism as an initial false start in search of 'total' regions, followed by less ambitious efforts to identify partial regions, this latter phase preceding and overlapping with the 'quantitative revolution'. The

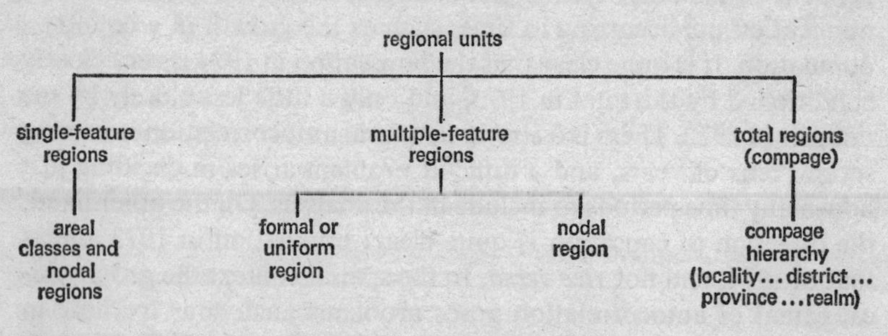

16. Categories of regions. (*Source: Haggett, 1965, p. 242*)

relationship of 'total' regions to Hegelian philosophical traditions has already been discussed (pp. 32–4), but efforts to identify them continued into the 1930s (Roxby, 1926; Passarge, 1929; Unstead, 1933; Stevens, 1939). However, doubts were being expressed. Roxby, for example, distinguished between the intrinsic conditions within a region and the external space relationships, which are unique and changeable for each region. Unstead went further, taking the view that in the last analysis all science exists only in the human mind and that therefore 'we may reasonably divide the earth's surface in whatever way is most useful at any given stage of our work' (Unstead, 1933, p. 175). By 1950, Wooldridge had arrived at the conclusion that geographers 'had better . . . abandon pseudo-organic analogies and we must recognize that there are many generic types of region' (Wooldridge, 1950, p. 2). A year later, Kimble was even more emphatic in suggesting 'that regional geographers may perhaps be trying to put boundaries that do not exist around areas that do not matter' (Kimble, 1951, p. 159). Thus did interest in 'total' regions wane.

Contemporaneous with this process was the elaboration of techniques to allow the identification of single-feature regions, a field of investigation that has remained fruitful to the present day. This is a topic that is discussed elsewhere in this book (see pp. 107–8). In the present section, attention is focused on the multiple-feature regions, which sub-divide into uniform and nodal regions. It is only in the last two or three decades that real progress has been made in devising suitable regionalization techniques for multiple-feature regions. However, before discussing these developments, some additional brief comments about the regional concept are in order.

First, the present author takes the view that regions are but a special case of the general problem of classifying data sets. From this viewpoint, regions do not exist as objective realities but as a convenient framework for description and analysis, and therefore the regional network will vary according as the purpose of the study changes. Second, in defining a regional system three considerations must be borne in mind:

(a) The kind of region and therefore the criteria for delimitation.

(b) The conversion of these criteria into an operational form to identify the regional boundaries.

(c) Whether the regional system is to be either or both: (i) space exhausting, and (ii) the regions mutually exclusive.

With respect to (c) above, it is usual, except when only one region is being considered, to assume that all places will be assigned to a region (a space-exhausting system) and simultaneously that no place may be located in more than one region (regions mutually exclusive), except in the case of hierarchically nested regional systems (p. 144). If these constraints are accepted, then it becomes very difficult to use complex criteria and operational forms under (a) and (b). This is a problem analogous to the difficulties encountered in transforming the earth's spheroid into a two-dimensional map, when the properties of correct shape, scale and area cannot all be retained simultaneously. If both the constraints of (c) above are to be satisfied, then it is usually impossible to respect multiple criteria for regional delimitation in a rigorous manner; something has to be sacrificed.

There are two basic approaches to the delimitation of multiple-feature regions. One may be described as the identification of boundaries, a matter that later is discussed as a problem in index construction (see p. 79). The other is essentially an assignment problem, or the grouping of areas. The former approach is now less popular than it used to be and most recent advances have been designed to achieve more efficient grouping procedures.

The grouping or assignment process starts from territorial blocks that are treated as given and for which relevant statistical information is available; parishes or local government units are typical examples. Given this array of *n* areas, the problem resolves itself into the following. What grouping of areas will yield a set of regions that, according to the criteria employed, have the least within-region variance and the maximum inter-regional differences? In formulating the question thus, there may or may not be a constraint imposed by the number of regions it is desired to identify; for example, both Berry (1961) and Spence (1968) have experimented with the number of regions into which the United States and England respectively can be divided. Also, for certain purposes such as defining administrative regions, it is usual to set a constraint that the areas grouped to make the region must be contiguous; if the purpose is description and analysis and not the definition of operational units, then the contiguity constraint may well not be necessary.

Haggett (1965) has succinctly reviewed the major techniques for the assignment of areas to regions and there is therefore little need to recapitulate his exposition. Suffice it to describe very briefly the nature of the basic principle, for which several operational forms are available. Imagine that there is an array of areas (say they are counties), for each of which information is available on a number of characteristics, as in Figure 4 (p. 28). If we consider one row, or vector, representing one characteristic, then it is obvious that according to this one criterion some counties will be 'near' to each other, and others will be far apart. For example, the population density of two counties may be very similar but dramatically different from other counties. Thus, each of the vectors can be regarded as a dimension along which the counties are separated by small or large distances. Hence, we may conceive of an *n*-dimensional space, *n* representing the number of vectors or categories of information. Each county then has an average distance (D) to each other county, obtained from the expression

$$D = \sqrt{\sum_{i=1}^{n} (x_i - y_i)^2}$$

where x and y represent pairs of counties and the subscript i signifies the values of one characteristic in the array 1 to n. Repeating the calculation for all pairs of areas, a table or matrix is generated showing the values of D. The pair of areas with the lowest value of D is then considered to be the pair with greatest similarity and therefore to be amalgamated as the first step in region building.

If we imagine four areas, A, B, C and D in Figure 17(a), then the

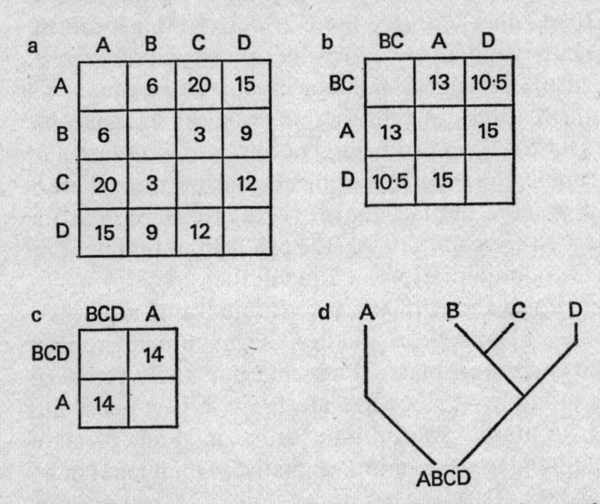

17. Stages in the grouping of a 4×4 distance matrix. (*Source: Haggett and Chorley, 1969, p. 245*)

distances that separate them can be entered in a 4 × 4 matrix. The lowest inter-area distance is between B and C, and these areas are therefore amalgamated to give a 3 × 3 matrix. It is then necessary to re-calculate some of the inter-area distances (Figure 17(b)) to identify which pair of areas is now closest together and therefore to be amalgamated to give two regions. The number of times that the operation is repeated depends on the initial number of areas and also upon the number it is desired to have at the end of the operation.

A grouping procedure in which average inter-area distances are calculated for an *n*-dimensional space involves three things of particular note. In the first place, the raw data may well be expressed in heterogeneous units and unless the data are first standardized some very peculiar results would emerge. For example, if one of the vectors were distance in kilometres and another population numbers, the grouping process would be drastically altered if the units of measurement were changed – for example, metres instead of kilometres. Consequently, the vectors are usually transformed and expressed as relative deviations from the mean value. Second, are all the characteristics to be given the same weight, or are some deemed more important than others? If so, there is a weighting problem similar to that encountered in index-number construction (see p. 79). In the third place, the essential problem is to make sense of a multi-dimensional situation by reducing it to more manageable proportions. The fact that this problem is not expressed in terms of traditional 'geographical' space (two-dimensional or three-dimensional) should not obscure the fact that it is still an exercise in the geometry of space with close affinities to the problem of representing the globe on a piece of paper. Expressed in another way, the major taxonomic problem up to about 1800 was to locate places accurately in terms of longitude and latitude; thereafter, the taxonomic problem was to identify like and unlike places, but using essentially verbal or qualitative modes of analysis; since the late 1950s, a new phase has opened up, based on more powerful and formal methods of taxonomy, that is, locating places (or, more generally, phenomena) in an *n*-dimensional space.

There is a further interesting parallel with the earlier problem of representing the globe in two dimensions. That problem consisted of eliminating one dimension, or vector. The modern equivalent goes by the name of factor analysis, the object of which is to collapse a large number of variables (the rows or columns in Figure 4) into a smaller number of dimensions, or factors. In the process, of course, some information is lost: more important, it is necessary that the research worker follow prescribed procedures and the user of the results understand the appropriate rules for interpreting them. This point has greater force the more complex the procedures that are adopted. It is salutary to remember how the political map of central Europe has been affected by successive and by no means always impartial attempts to map the ethnographic distributions (Wilkinson, 1951) and how the Nazi regime fostered the deliberately misleading representation of European geography (Whittlesey, undated). With this point in mind, Robson's comment on the use of factor analysis in the urban context applies to the technique in general:

> It might appear that factor analysis, ideally suited in many ways to tackle the problems of urban ecology, should have solved many of the questions posed by Social Area Analysis by providing clear indications of whether or not there are universal dimensions of urban social structure and, if there are, how these are distributed in space. The fact that the use of the technique seems not to have resolved such questions may be related both to the problems of making comparisons between cities and to certain technical and operational uncertainties of the method itself.
>
> (Robson, 1973a, p. 207)

It seems a safe prediction that geographers and others have enough work to last several decades to establish standard and generally accepted rules for the measurement of multi-dimensional space and the analysis of phenomena in this context.

SPATIAL INDEX NUMBERS

For quite some time now, geographers have been in the habit of using indices to represent spatially complex data, but for the most part have been apparently unaware that the indices employed are in fact index numbers of one kind or another. Most of the literature concerning the theory of index numbers has been developed in the temporal rather than the spatial context and therefore it is useful to begin with a brief recapitulation of the basic ideas.

The classical domain for index-number construction is that of price changes over time. Faced with numerous commodities, each changing in price at a different rate, it is convenient to have a single figure showing the overall change, such as the familiar cost-of-living index. The fundamental principle involved is to obtain the unit price for each commodity at two points in time. A weighted average figure is then derived by assigning each commodity a weight, representing its relative importance in the bundle of items being considered. Thus, the two basic forms of index number are given by the following expressions:

Laspeyres index $$I_n = \frac{\Sigma P_n \,.\, Q_o}{\Sigma P_o \,.\, Q_o}$$

Paasche index $$I_n = \frac{\Sigma P_n \,.\, Q_n}{\Sigma P_o \,.\, Q_n}$$

where:

I = price index
P = unit price of a commodity
Q = quantity consumed of a commodity, or its weight

and the subscripts *n* and *o* denote respectively the given year and the base year. The Laspeyres index uses base year weights and shows what the overall price would be in the given year if consumption patterns remain unchanged; the Paasche index uses the given year weights. (For further details, see Allen, 1966.)

Now, instead of comparing base year and given year it is simple to substitute base region and the region in which one is interested for comparison with that base. At the time of the enlargement of the European Economic Community (effective 1 January 1973), there was intense interest in Britain in just such price comparison. Similarly, the United Nations has for some time maintained indices of living costs in the world's capitals as the basis on which to adjust the remuneration of its officials.

The classical problems in index-number construction can be stated quite briefly. First is the decision as to which items to include, a decision that is necessarily partly subjective but usually based on a judgement of their importance in the system under consideration. Second is the question of which weights to use – base year (region) or 'end' year (region), or the average of the two. In the temporal context of rapid change in the weights, it is also possible to construct the index on a changing weight basis by pairwise linkage of time periods; there seems to be no feasible spatial equivalent of this procedure. Third and last is the universal problem of the homogeneity or otherwise of the items being considered.

There are two main families of spatial index numbers, which may be described respectively as measures of structure and of components of change. Chisholm and Oeppen (1973) have reviewed the development of the first category, with special reference to industry, for which general interest was only aroused with Florence's 1944 paper on the dispersal of industry in Britain and Weaver's 1954 study of crop-combination regions in the United States. Note, however, the earlier rather qualitative work of Jonasson (1925) on agricultural regions and the attempt by Hartshorne and Dicken (1935) to frame a quantitative classification of farming areas, as well as Hoover's (1936) work.

Measures of structure may be sub-divided into two classes. The first is exemplified by Florence's approach to industrial localization, from which a measure of the specialization of employment in regions may be derived. Given total employment for a region for each of n industries and for a 'base' area – say the national totals of employment – the absolute values can be converted to percentage terms. If the national percentage values are subtracted from the regional percentages, the positive (or negative) deviations can be summed and then for convenience divided by 100. The index of specialization so derived has a range from zero, when the regional distribution of employment is identical to the base distribution, to approaching 1·0 when employment is concentrated in a single industry. The second category of indices of structure is based on some ideal distribution, with which the actual distribution is compared. Weaver's method of identifying crop-combination regions compares the actual percentage distribution of crop areas with the ideals of one-crop (1×100 per cent), two-crop (2×50 per cent), three-crop ($3 \times 33{\cdot}3$ per cent), etc. combinations. The generalized idea is that of Lorenz curve analysis, as illustrated in Figure 18. If the categories represent different land uses, for example, then a uniform distribution among the classes would be represented by the straight line (a) showing the accumulated percentage. Where the distribution is unequal, and if the land-use categories are ranked from highest percentage to lowest, then curves, such as (b) and (c), may be obtained. In Figure 18, curve (c) represents a high degree of concentration into a few land-use categories, whereas (b) describes a more diverse pattern of land uses. Various ways have been devised to provide a convenient numerical description of these Lorenz curves, so that differing distributions may be compared (see Chisholm and Oeppen, 1973; Yeates, 1968).

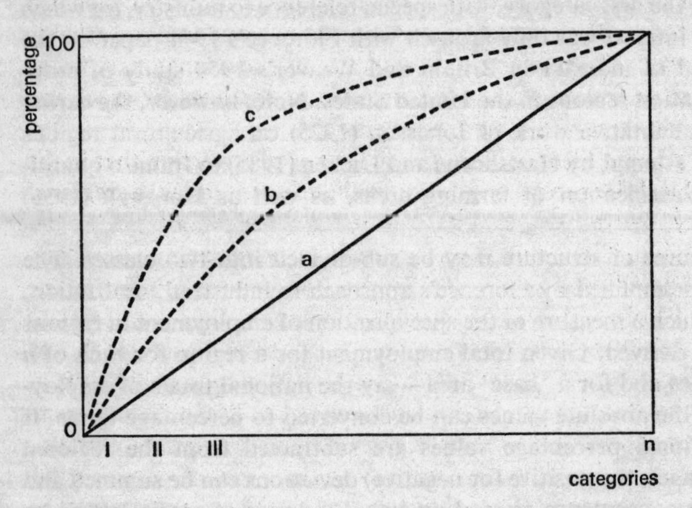

18. Lorenz curves.

There has been surprisingly little discussion in the geographical literature of the merits and shortcomings of the various indices. Without a clear recognition of these properties, it is only too easy to use them inappropriately. This point has been made by the present author (Chisholm, 1964) concerning the land-use classification devised by Hartshorne and Dicken (1935); their classification was based on patterns of land use and therefore did not permit of conclusions about farming systems, since the use to which products were put was ignored. This was a mistake that Coppock (1964, 1971) carefully avoided in his regionalization of British agriculture on the basis of standard labour requirements for crop and livestock enterprises. Note also that Hoag (1969) has shown how Weaver's method for identifying crop-combination regions – a method which can of course be extended to industry (Johnson and Teufner, 1968) and other distributions – is subject to potential error and confusion.

Ferguson and Forer (1973) addressed themselves to the problem of whether one index of employment specialization yields results that are similar to other measures, considering especially the effects of the population totals of the geographical areas used for statistical purposes and also the level of industrial aggregation. They concluded that the four indices they selected for study behaved in a remarkably similar way, indicating that the choice of index may not be as important an issue as the need for care in the level of areal and category disaggregation used in a study.

An important form of spatial index is the linear version that in the geographical literature is traditionally referred to as a regional boundary, derived as a synthetic value from several criteria. To delimit a region, it is common practice to superimpose several boundaries, each based on a single criterion, and to derive a composite line as some average value. Each criterion is normally given equal weight with all the others though there is no intrinsic reason why this should be so (Haggett, 1965, pp. 245–7). Other aspects of regional identification are treated on pp. 76–8 and therefore suffice it here to note that regional boundaries are but a special kind of geographical index.

To measure the components of change in a spatial system, shift-and-share techniques became popular in the 1970s and were subjected to a good deal of criticism and development (see Chalmers, 1971; Bishop and Simpson, 1972). The fundamental idea of shift-and-share analysis is readily stated. Given a system of regions, some experience a faster rate of employment growth than others and, indeed, in some regions growth may be negative. A problem arises in determining how much of the *difference* between actual regional growth and the national average can be attributed to the two factors:

1. The rate of growth of the separate industries in the regions.
2. The employment structure of the regions.

To establish the relative importance of these two components, a simple principle is adopted. For any given region, the employment structure is known for two points in time, and hence the rate of change of each industry. If the regional employment structure in the initial year is taken as the datum (equivalent to using base year weights), the relevant national rates of growth can be applied to these employment totals, to obtain the expected overall performance of the region on the assumption that industries grew at the national rate. For each industry in each region one then has a measure of the extent to which it is actually doing better or worse than the same industry is doing nationally. Hence, it is possible to distinguish how much of the region's relative growth performance is due to the rate of growth of the industries and how much to the preponderance in the region of nationally fast- and slow-growing industries.

The technique involves problems already noted in the construction of index numbers, notably whether to use initial year or terminal year weights and the level of aggregation to be employed. While useful as a descriptive tool, its analytical power seems to be limited on account of the correlation of growth rates both between industries (industrial linkage) and between regions (spatial autocorrelation).

In sum, spatial index numbers are of considerable interest as yet another way of dealing with the complexities of multi-dimensional space. Though historically index numbers have been developed largely independently of factor analysis and similar techniques, they share the property of reducing many dimensions to few.

ABSTRACT SURFACES

So far, we have been concerned with the various ways in which 'real' phenomena can be represented, including the idea of constructing generalized surfaces to replicate variations in population density, income levels, etc. In all the approaches to description so far discussed it is implicit that some concrete, almost tangible, phenomenon is being described. A much more abstract concept was introduced after the last war by American scholars (Stewart, 1947; Stewart and Warntz, 1958), that for convenience may be termed surfaces of potential interaction. The fundamental idea is closely linked to the concept of gravity, and may be stated quite simply. Imagine that you live in New York. Within a very few kilometres many millions of other people also reside and you therefore have very many people with whom you may potentially come in contact in the course of work and leisure activities. You might also meet inhabitants from Chicago, either when they visit New York or you travel westwards: similarly, you might encounter any of the inhabitants of any city in the United States. The chances of meeting someone – anyone – from a particular city will clearly vary for two reasons: the bigger the other city and the nearer it is the greater the chances that you will meet someone from it within a specified time period. Stated more formally, the probabilities of an encounter are proportional to city size and inversely to the intervening distance. Thus, the *potential* accessibility of any one place to all other places in the United States can be obtained as the summation of city populations divided by the respective distances to them. The larger the summation, the greater the contact potential, and *vice versa.* The resulting map is generally described as a map of population potentials.

Population is only one of the variables for which a generalized map of potential accessibility may be wanted. As a salesman for a large firm, it would be useful to know how accessible all locations are to the total retail sales of the country, or sales of a particular commodity, and so maps of income or sales potentials could be constructed on the same principle as for population potential (Figure 19). These are the main uses to which abstract surfaces have in fact been put, but the reader will recognize that the idea could be used in a variety of other ways; Warntz (1959), for example, compiled maps of supply potential for selected crops in the United States as part of his study of the spatial variation in agricultural commodity prices.

REGULARITIES ABSTRACTED FROM TWO-DIMENSIONAL SPACE

Imagine a globe representing the world, on which are placed symbols to locate the major cities and represent their respective populations. Such a globe effectively uses four dimensions – three to describe the

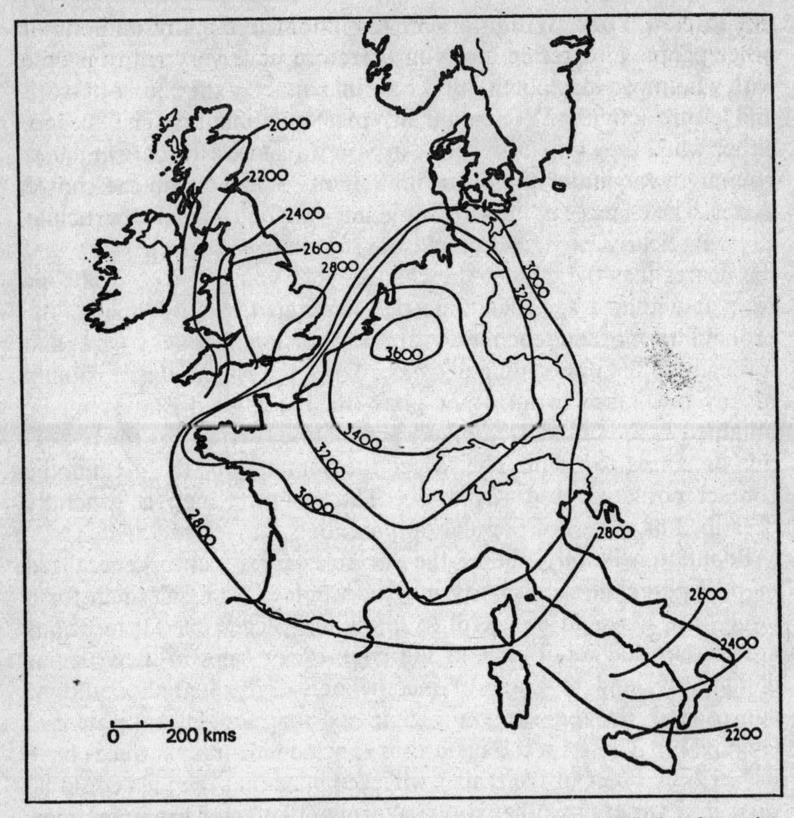

19. Income potentials for the European Economic Community and Britain prior to the latter's accession in 1973. (*Source: Clark et al., 1969, p. 204*)

positions of the cities and one their size. If one dimension is eliminated, the size of cities could be shown on a map. That two-dimensional map could be further simplified if a single city were chosen as the point of reference and all other cities were plotted to show their size in relation to distance from the origin. If even that one space vector were eliminated, city sizes could be represented as in Figure 20 for the United States. The horizontal axis shows cities ranked from number one, the largest, to the smallest; the vertical axis shows the population total. Thus, the curves show the relationship between city rank and population, a relationship that in the United States is remarkably consistent at each date and has changed over time with remarkable regularity.

Since Singer (1936) drew attention to the rank–size rule, economists, geographers and others have been fascinated by these regularities, both to determine statistically the best descriptions and to unravel the processes that have given rise to them (for example,

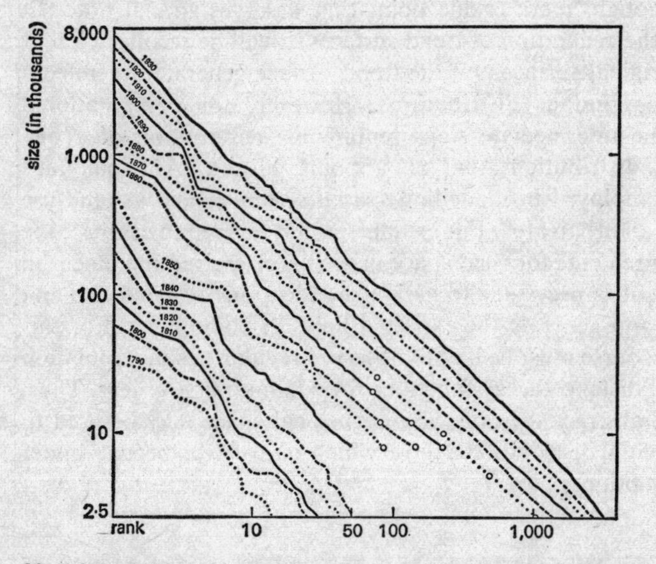

20. U.S.A., 1790–1930. Communities of 2,500 or more inhabitants ranked in decreasing order of population size. (*Source: Zipf, 1949, p. 420*)

Richardson, 1973a and b). This interest was spurred by Zipf's seminal book *Human Behavior and the Principle of Least Effort* published in 1949 and yet further increased by Clark's (1951) discovery that similar statistical regularities exist for the density of population in a city – excluding the central area, density decreases with distance from the centre (see Mills, 1970, for a most useful summary of work in this field). It quickly became apparent that many phenomena can be expressed as remarkably consistent regularities in just two dimensions, one of which may be a distance vector: migration flows, telephone calls, commuter trips – all show pronounced distance-decay features, that is, an inverse relationship between distance and the amount of interaction. These well-established regularities are summarized in several works (Isard, 1956; Haggett, 1965; Olsson, 1965; Berry, 1967). Although there is still doubt why these regularities occur, some writers dubbing them as facts in search of a theory, a start has been made in explaining the logical structures that underlie them (see Chapter 5). For the present purpose, though, there is one point that deserves special note. In discussing the generation of trend surfaces, it will be recollected that both the original surface and the trend surface generated therefrom represent continuous distributions. However, for computational purposes the intermediate steps require the assumption of a discontinuous distribution, that is, a point pattern. An analogous process is employed in modelling spatial systems, such as land-use patterns in an urban area. For example, if it is desired to estimate the future requirements for road space in a city for a given increment in population, it is necessary to assign the locations of residence and workplace and generate the expected flows of commuters. For this purpose, recourse must be had to observed regularities of population density and distance-decay functions for commuting journeys. Thus, regularities abstracted from two-dimensional space may be used to re-create spatial distributions, those which are to be 'expected' under given assumptions.

OTHER PROBLEMS OF PATTERN DESCRIPTION

'Shape is one of the most difficult properties of a geographic pattern to measure' (Haggett, 1965, p. 228). Some geographers have examined the problem of devising standard measures of shape that can be applied to administrative units, shape of settlements and similar matters (Bunge, 1966), but progress has in fact not been great. There may be two reasons for this apparently unsatisfactory state of affairs. In the first place, there is a large number of shapes that may be regarded as norms and there are no obvious guideposts to indicate which is the best approximation to a given real shape. Second, where ratios of perimeter to area, longest to shortest axis, etc., are employed there is at once an uncomfortable degree of subjectivity involved in the measurements and at the same time a serious lack of theory to which the measures can be related, at least in the field of human geography. Thus, while physical geographers have a reasonably clear set of reasons for being interested in the shape of soil particles and pebbles, it may well be the case that in human geography attempts to measure shape have been premature.

On a more hopeful note, one of the incidental benefits to be derived from the development of formal methods for describing spatial patterns lies in the field of sampling. Some kinds of data cannot be obtained from censuses or similar sources which enumerate the total population of the phenomena in question. For the researcher to undertake a complete tabulation is usually an impossible task. Therefore, the research worker must very often draw from the universe a sample for his study. In the absence of proper techniques of spatial analysis, such samples cannot be guaranteed to be either random or correctly stratified. Until the end of the 1950s, the techniques currently available did not permit full advantage to be taken of sample design in a spatial framework. Now, however, it is much more practicable to design samples that will enable spatial patterns to be described economically and accurately (Gregory, 1963; Gould, 1969a).

CONCLUSION

The reader will have noticed that for many of the descriptive techniques mentioned in this chapter general acceptance and use by geographers came only after the Second World War. In this sense, the rate of change has indeed been dramatic. However, by using prose as the medium of exposition and by concentrating on the essential underlying ideas, it is manifest that continuity of thought dominates over the apparent discontinuity which some have called the ‘quantitative revolution’. An analogy may make this point clearer. In survey work the fundamental principles are derived from elementary trigonometry and are conceptually so simple that most children can grasp them. However, to make these principles operational in a precise manner requires the use of complex equipment to provide the basic data. More tiresome, though, is the fact that in ground survey errors are introduced by the refraction of light through layers of air of varying temperature; thus, theodolite observations for the base triangulation are normally restricted to the few hours after dawn. More complex still was the problem encountered in the survey of India: errors crept into the altitudinal calculations because of the then unsuspected effect of gravity anomalies. Thus, many of the complexities of practical surveying arise from attempts to eliminate error, not from the difficulties of the underlying principles. In terms of this analogy, much of the recent work in the identification of spatial patterns can be viewed as attempts to reduce error and to specify the magnitude and possible nature of that error which cannot be eliminated.

At various points in the preceding pages, attention has been drawn to the similarity in principle between the various transformation operations and the task of representing the globe on a flat piece of paper. Herein lies a major element of continuity in our subject, namely, the traditional interest in geometry. It is the geometry of space with which this chapter has been concerned, and, because space may be conceived in many ways, many geometries are involved. But the fundamental link between the descriptive techniques discussed is the difficult twin problem: how to describe the locations in *n*-dimensional space of one phenomenon relative to another; how to collapse many dimensions to few. Arising out of these issues is the further one: how to relate one pattern to another, as part of an explanatory system, whether this system be inductively or deductively derived. Most important of all, though, the geometries that we use and our related conceptions of space are not independent of the purpose of our inquiry. The two are intimately related, firstly because the way we interpret patterns depends on our understanding of the logical principles underlying the data-transformation process, and secondly because pattern recognition and theoretical explanation are but two faces of one coin. To develop a theory, some metric of space must be used; to test it, some transformation of that metric must be made to another that can be made operational with the data at our disposal.

As a final point that has been implicit throughout the discussion, but has not hitherto been made explicit, the problems of description with which this chapter has been concerned apply at all geographical scales, though in varying degree. It is not useful to think of particular techniques as relevant only to specific scales of inquiry, whether these be the spatial pattern of cropping within a farm, intra-urban distributions or national/world-wide patterns.

4. Dynamic Patterns

Thus far, we have reviewed the problems of accurately describing spatial patterns of a static nature and we have seen that in recent decades there have been very substantial advances, building on the work of many generations. The time has now come to examine the issues at stake in describing dynamic patterns. Several of these issues have been admirably stated by Abler, Adams and Gould and one can hardly do better than quote them *in extenso*:

> Contemporary geographers devote a great deal of attention to spatial structures of all kinds. Almost any substantive problem a geographer tackles can fruitfully be considered to be a problem of describing accurately and explaining satisfactorily the spatial structure of a distribution. The emphasis in contemporary geography on spatial structure is somewhat misleading, because it overemphasizes distributions to the neglect of the *spatial processes* which interact causally with them. What we call spatial processes are mechanisms which produce the spatial structures of distributions. Reference to spatial process is inescapable in any explanation of spatial structure.
>
> Why geographers are so much more aware of distributions than of the processes which produce them is not clear; it is probably because distributions of static things are easier to observe and record on maps than the processes which produce them. Their view that spatial structure is primary in their science often leads geographers to overlook important relationships, not the least of which is the causal effect of structure on process. Spatial structure and spatial process are *circularly* causal. Structure is a determinant of process as much as process is a determinant of structure. The existing distribution of supermarkets in a city, for example, is a fundamental determinant of the success of any new supermarkets established in the area.
>
> In proper perspective, the distinctions we make between spatial process

and spatial structure disappear because they are based upon a limited time perspective and are thus somewhat artificial. Processes are spatially variable and thus have distributions just as do 'concrete' phenomena. Thus spatial structure is a concept applicable to both static distributions and to processes which appear to us to be dynamic. Process and structure are, in essence, the same thing. Whether we see process or structure when we look at a spatial distribution depends on the time perspective we adopt and the rapidity with which the process moves. Human movements in vehicles and on foot result in spatial structures of objects like roads, railroads, airports, and sidewalks on the surface of the earth. Similarly, human choices of agricultural, industrial, and commercial activities produce economic spatial structures.

Because we find it easier to make maps of distributions of apparently static physical phenomena than of moving people, or their motives for moving or for making the choices they do, we tend to think that these distributions of physical phenomena are as static as they appear to be on maps. Yet upon reflection it is obvious that distributions and their structures constantly change, and when we consider apparently static distributions over periods of twenty-five or fifty years, their structures are usually very dynamic phenomena. Imagine, for example, a series of weekly maps of railroads in service in the United States and Canada over a period of fifty years. If a film were made of this sequence of 2,609 maps which showed twenty maps per second, the network would contract in some areas and expand in others in a very dynamic manner.

It is our limited time perspective which makes us designate some spatially differentiated phenomena as distributions of objects while we designate other, more rapidly changing distributions as processes. When we distinguish spatial process from spatial structure we are merely recognizing a difference in relative rapidity of change; both the obviously changing and the apparently static distributions on the earth's surface are components of spatial processes. 'Spatial distribution' is a term we apply to spatial processes which appear to us to be static, and 'spatial structure' is the term we apply to the internal spatial organization of these distributions of process elements. Properly considered, the spatial structure of a distribution is viewed as an index of the present state of an ongoing process.

(Abler, Adams and Gould, 1971, pp. 60–61)

Three ideas in the above passage deserve special attention. The first and the most obvious is that the distinction between static and dynamic situations is essentially a matter of the time-period adopted for the study. Their idea of constructing a film from weekly maps of the United States railway system is a particularly vivid illustration of this notion, which clearly has a general application to phenomena of all kinds – changes in population distribution, income distribution and trade flows would be some examples that come readily to mind. The second idea is no less important. A phenomenon such as a road network, which in the short run may be regarded as a static pattern with a very real geographical expression, provides a facility for the *movement* of goods and people. In other words, there is a highly dynamic pattern of flows along the network elements. Furthermore, just as the flows 'create' the network, the very form of that network conditions the nature of the traffic movements that can be accommodated. Consequently, the 'static' features on the earth's surface are, in a very real sense, surrogates for activities or processes of all kinds. However, underlying these observations is the difficulty, indeed frequent impossibility, of inferring the relevant processes from a pattern true for a single moment of time. Hence, the whole thrust of the passage quoted above is the need to study the processes which operate to create the geographical patterns we observe. In this way, we may hope to learn not only how things come to be as they are but also how they are likely to unfold in the future.

In approaching the study of processes more directly than is attempted in the previous chapter, it would seem logical to consider first the scope for *direct* measurement of what is going on. However, there are two reasons why this course of action has not been adopted. As Abler, Adams and Gould have observed, in the historical development of geography as a discipline, attention has been directed mainly to spatial structures, Broadly speaking, the next step was to arrange the spatial structures in a time sequence so that inferences about processes could be drawn. Only with the advent of quantification has the measurement of processes become a central part of geographical work (see p. 44). Therefore, to adhere to the purpose of the present book, which is to discuss the way geography has changed and continues to change, it is most advantageous to proceed as follows. The next section discusses ways of analysing networks as surrogates for the movement of goods, people and information. If you like, the networks are the vehicles for the transfer of 'mass' or 'energy' and the form of these networks, and their changes over time, allow some inferences to be made about the activity patterns. Next follows consideration of the more direct study of flow patterns, followed by the role of matrix methods, movement regularities abstracted from two-dimensional space and space–time budgets. This last topic represents one of the newer departures in geographical description. It might be logical to continue with the studies of environmental perception and decision-making procedures but other considerations suggest that these topics are best left to Chapter 5. The present chapter, therefore, continues with a brief review of historical geography, a branch of the subject established long before many modern analytical techniques were invented but which displays great vigour right to the present day.

NETWORKS

It was not until 1936 that König published the first comprehensive treatment of networks. This book is regarded by Haggett and Chorley as being of seminal importance and there can be no doubt that serious study of the properties of networks largely dates from its publication. This fact is clear from the pattern of references in Haggett and Chorley's *Network Analysis in Geography*, published in 1969; of the 430-odd items of literature to which they refer, only about 120 antedate 1960 and of these the vast majority were published in the 1950s. Network analysis has until recently been something of a novelty, both in geography and other disciplines.

The rudiments of network analysis have already been indicated by Figure 10 (p. 59), showing how a network may be both simplified and described in different ways. Referring to 10(b), it is easy to identify the nodes (or junction points), the links between these nodes and the areas that are partially or totally enclosed by the links. It is also intuitively apparent that there may be some interesting relationships between these three network elements. However, before we consider these, it is useful to distinguish two kinds of network. The simplest is a branching network, such as a tree (Figure 3, p. 21): the branches sub-divide into smaller ones and ultimately twigs, each twig being a terminal point or dead end. For geographical studies, the most important class of branching network is provided by rivers. On the other hand, Figure 10 represents a circuit network, as exemplified by road systems and airline networks. This basic distinction is important because the properties of the two network types differ, as will become apparent presently.

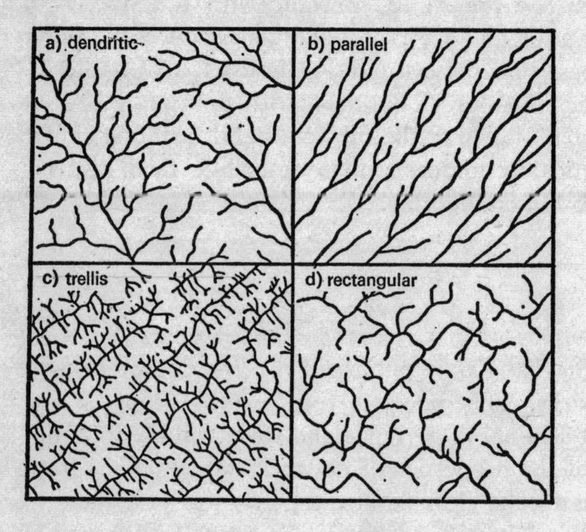

21. Basic types of drainage pattern. (*Source: Haggett and Chorley, 1969, p. 90*)

Geomorphologists are specially interested in the branching networks of river systems and their relationship to both terrain and climate. The reader will recollect that point patterns can be described as clustered, regular or random (p. 68); an analogous categorization of networks may be made, only this time on a four-fold basis – into dendritic, parallel, trellis and rectangular patterns (Figure 21). The description can be amplified by including information on the orientation of the network segments, as north–south, east–west, etc. Perhaps more important, though, are other network characteristics.

Different segments of a stream system exhibit different morphometric and hydrologic features and relationships, and earth scientists are particularly concerned with their hierarchical organization. This has required the assignment of a level of relative order of magnitude to each segment in a stream network hierarchy, determined by the sequential arrangement of tributaries with respect to the main trunk ... As Strahler has pointed out, practical utility is the criterion by which the success of stream-ordering techniques must be judged, for any usefulness which the stream order system may have depends upon the premise that on average, if a sufficiently large sample is treated, order number is directly proportional to relative watershed conditions, channel size and stream discharge at that place in the system.

(Haggett and Chorley, 1969, pp. 10–11, in part quoting Strahler)

The remarkable nature of the regularities that can be identified using the concept of stream order is vividly shown in Figure 22. Data for this diagram were obtained for a comparatively small

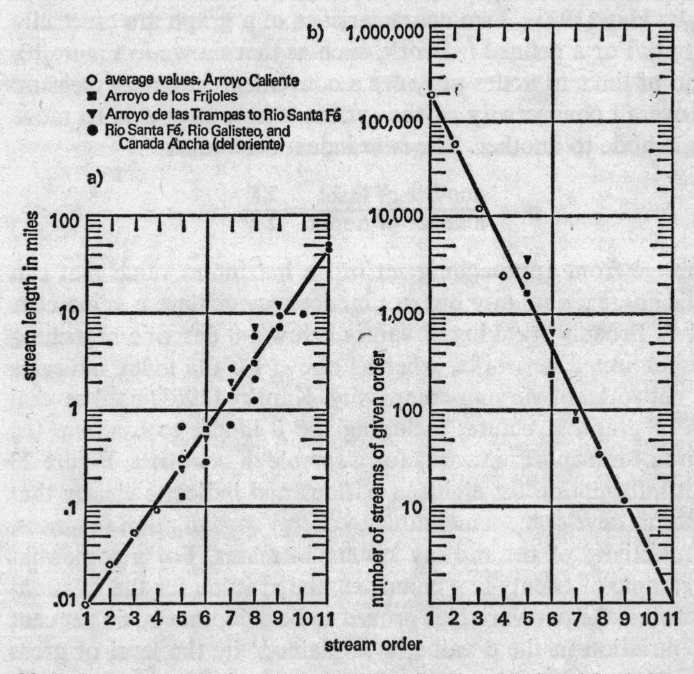

22. Drainage network of arroyos near Santa Fé, New Mexico. (a) relation of stream length to stream order; (b) relation of number of streams to stream order. (*Source: Leopold et al., 1964, p. 140*)

drainage basin in New Mexico. However, as Morisawa (1968, p. 154) points out, most river basins appear to exhibit similar regularities and the stage has now been reached at which one may fairly say that empirical 'laws' have been identified, laws for which geomorphologists are now seeking explanations.

For human geographers interest focuses on circuit networks. The most tangible are transport facilities, such as road, railway and canal systems; airline and shipping networks are not visible features on the ground, except for the nodes (airports and seaports). The study by geographers of transport systems in network or graph terms only started with Garrison's 1960 paper on the connectivity of the American inter-state road network (Gould, 1969a, p. 16). Since then, a great deal has been done, as witness the work collated by Haggett and Chorley in *Network Analysis in Geography* and the more recent review by Hay (1973). Two characteristics of a graph are especially interesting. For a defined network, such as that shown in Figure 10, the ratio of links to nodes provides a convenient summary measure of the overall connectivity of the system – how easy it is to move from one node to another. The *beta* index, derived as

$$\beta = \frac{\text{number of links}}{\text{number of nodes}} = \frac{\Sigma L}{\Sigma N}$$

has a range from approaching zero to a maximum value that can theoretically reach infinity but in practice may achieve a value near to 2 or 3. Broadly speaking, β values below 1·0 describe branching and disconnected networks, whereas above 1·0 the index indicates circuit networks of varying complexity. Kansky (1963) used several indices of graph structure, including the β index, to examine the structure of transport networks for a sample of countries. Figure 23 sets out information for eighteen nations and indicates clearly that as economic development measured by energy consumption advances, the connectivity of the railway system increases. For a somewhat larger sample of twenty-five countries, the r^2 value for the relationship between the two variables proved to be 0·64, that is, 64 per cent of the variation in the β value is 'explained' by the level of gross energy consumption.

Instead of considering the whole network, attention may be

directed to individual nodes and the question asked: How well is each node connected to all the others? At the most elementary level, it is easy to see from Figure 10(b) that to get from node 1 to node 4 requires the traverse of only one link, whereas to reach node 17 it would be necessary to use seven links. In principle, it is simple to calculate the average number of links from any one node to all the

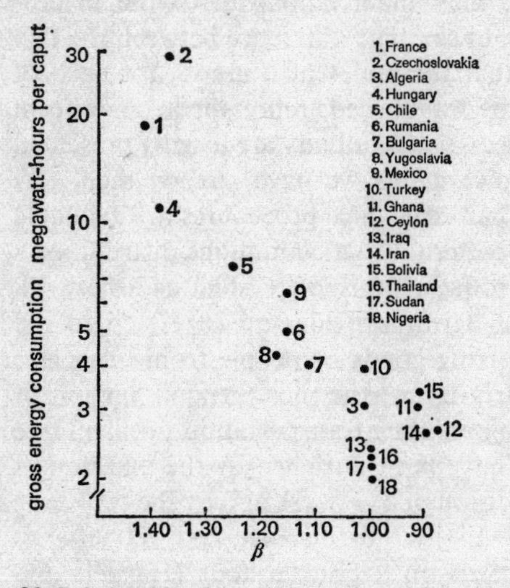

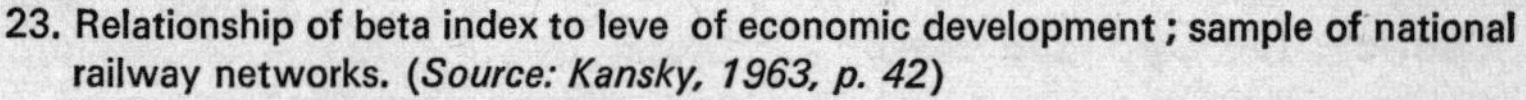

23. Relationship of beta index to leve of economic development; sample of national railway networks. (*Source: Kansky, 1963, p. 42*)

others in the network; the lower this average, the more highly connected the node. Hence, comparison may be made between nodes within the system and also one may examine the changing absolute and relative connectivity of places as the network evolves.

Other indices of connectivity have been devised. However, there is a difficult and still only imperfectly solved problem of weighting links and nodes to take account of route length, transport costs, importance of the nodes, etc. Furthermore, graphs (or networks) can be sub-divided not only into branching and circuit categories but also according as they are or are not 'planar' graphs. A planar graph has no intersections between edges except at nodes, whereas for non-planar graphs this condition does not hold. Notwithstanding these elaborations, the reader will readily recognize two characteristics of network analysis.

The idea of connectivity in a graph is essentially the same as the idea of potential accessibility developed by Stewart and Warntz, which has been discussed in Chapter 3. If connectivity in a graph is conceived in terms of cost or time instead of the more primitive notion of number of links, then the relationship to the macro-geography approach is easier to see. The difference between the two approaches lies in the fact that the nodes of a graph, for each of which the connectivity can be established, represent a punctiform pattern, whereas maps of potential accessibility are usually presented as a continuous surface. However, as we have already seen, this difference is more apparent than real, since procedures are available for converting a punctiform pattern into a continuous distribution.

In the second place, regarding a transport system as a network that can be treated in graph terms provides an easy step to the problem of efficiently transferring goods or people from one place to another. Reduced to matrix terms, the most general version of this problem is the linear programming transportation problem (see Chapter 5), though workers have been intrigued by the question of how to find the shortest path through a network and by the travelling salesman's need to minimize his journey distance and yet visit all of a specified number of places on a circular tour (Haggett and Chorley, 1969, pp. 199–204).

FLOW PATTERNS

Although transport had for long been recognized as an essential element in geographical studies, and though a modest literature had appeared in Germany, France, Britain and elsewhere, relatively little progress had been made before the early 1950s. The greater part of the then available geographical publications were confined to descriptions of route patterns, economic considerations such as rate structures and patterns of international trade. Ullman identified the main reason why the actual transactions, or flows, had been little studied:

Transportation geography, due to the lack of proper data, is still in the stage of searching for better methods of measurement and mapping. Many old conclusions and concepts are unreliable because they are based on insufficient quantitative data, such as the erroneous idea that the New York Central, rather than the Pennsylvania [railway], has the heaviest traffic between east and west.

(Ullman, 1954, p. 317)

In terms of data availability, the study of transport geography was at least fifty years behind other branches of the subject. Hence the need to rely on surrogate measures for spatial interaction, such as Ullman's 'map of the railroads of the United States [which] classifies lines according to number of tracks and type of signalling, the best quantitative indicators of flow and of relative importance available from records' (Ullman, 1954, p. 317).

Ullman's remarks related specifically to the movement of goods but applied with equal if not greater force to other forms of interaction. Patterns of commodity flow are directly related to the localization of productive activities, comparative advantage and regional specialization (Ohlin, 1933). At the international level, data on commodity values in trade have been readily available for many years, as also the number of shipping movements. However, quantities are recorded by miscellaneous units – tons of wheat, barrels of oil and square metres of cloth, etc. For example, only since 1965 have figures been available regularly for the tonnage of all goods recorded as imports to and exports from the United Kingdom, but, even now, these data do not allow the national origins and destinations of the tonnages to be identified. At the intranational level, there has until very recently been little or no information on freight movements except on a sample basis. Only in 1947 did the United States Commerce Commission start to publish an annual one per cent sample of rail waybills (Ullman, 1954, p. 317), at about the time that origin–destination surveys of road traffic began to be undertaken in that country, mainly as an aid to urban planning. Great Britain lagged behind; not until 1964 was the first thorough estimate made of inter-regional freight traffic moving by road and rail (Chisholm and O'Sullivan, 1973) and it appears that data of similar quality may not be prepared again. Ironically, some of the less developed countries are better documented, at least for some aspects of freight movement; the Nigerian railway system is a case in point (Hay and Smith, 1970).

Rather similar data problems arise in documenting the permanent migration of people. Relatively good statistics exist at the international level but there are few countries that can rival Sweden for the detail and time coverage of its internal population records. Whereas in Sweden the annual movement of individuals can be traced from parish records, the position in Britain is that to establish the main long-term migration flows requires careful inferences from decennial censuses, in which information is available only for present residence and place of birth (Friedlander and Roshier, 1966). The position has been somewhat improved with the inclusion in the more recent British censuses of questions concerning place of residence one and five years previous to the date of enumeration, but as the identity of individuals is not linked from one census to the next there is no possibility of recording the life-time migration patterns of individuals, except by specially designed sample investigations. The extent of our ignorance about internal migration is dramatically revealed by the fact that in February 1968 the Department of Employment *cancelled* data for the inter-regional movements of workers that had been previously published: it was found that the basis for estimation was yielding highly unreliable figures (*Employment and Productivity Gazette*, LXXVI, 1968, p. 120).

The daily flow of people to and from work began to be documented only in the present century. Germany was the first country to include in a census questions on commuting, in the year 1900 (Liepmann, 1944, p. 111), followed by Switzerland in 1910 and Britain in 1921. Not until after the Second World War did it become a regular feature of national censuses to collect information on commuting. Thus, while Lawton (1963 and 1968) was able to use comprehensive data on the journey to work, Liepmann in her earlier and pioneering study had had to be content with piecing together some census evidence and the results of special inquiries. Altogether, therefore, it is only in the very recent past that adequate documentation in this important aspect of life has become generally available.

As with the movement of goods and people, so with capital and current-account flows of money. Reasonable international statistics have existed for quite some time, but intranational records are usually rudimentary or non-existent. Japan is uncommonly well documented in this respect, the Bank of Japan collecting statistics on government cash transactions (tax receipts and expenditure) as an inter-regional matrix for eight regions comprising the whole country (Chisholm, 1966, pp. 217–19). National clearing banks must have records showing the inter-regional movements of private funds, but these data are not normally available for analysis. Given the problem of confidentiality it seems unlikely that much progress can be made in this sphere, though the importance that attaches to confidentiality varies from one nation to another. For example, I understand that records are available in India for the cash flows between post office branches, flows that are relevant as a measure of small private transfers, as between members of the same family.

The fourth and last kind of flow with which we are concerned is the movement of information and ideas. At the time Hägerstrand prepared his seminal book on the diffusion of innovations, published in 1953, he had no directly measured evidence of the way in which ideas were transmitted. Subsequently, Swedish workers in particular have mounted an intensive programme of research to record the spatial contact patterns of businessmen, government officials and others (Törnqvist, 1968), and similar work, closely influenced by the Swedes, is being undertaken by Goddard (1973) among others.

Poverty of data on flow phenomena is not as surprising as may at first sight appear. To describe the location of an object requires one pair of co-ordinates, and the object in question can be recorded at a time which suits the observer – whether a surveyor from the national survey organization or the geographer doing field work. Accurately to record a flow, however, is much more complex. At the very least, both the origin and destination must be identified, and for preference some information on the route followed as well. Furthermore, as flows may occur at any time in the 8,760 hours that comprise a year, a single observation will hardly suffice and some special system of data recording must be established. As a general principle, it is only in a literate society that the expertise exists to do this. Similarly, it is only where there is a high degree of interdependence between parts of the economy that there is a need for such data. Thus, it is really only in the present century that the need has emerged for intranational flow data over a wide front and hence that statistics have begun to become available. That said, there remain formidable problems in providing an accurate description of flows, if description is to do more than record the volume of traffic that passes defined points in the network, for example, the records of traffic volume for selected points on Britain's roads. Two problems in particular deserve mention.

Discussing the nature of road freight flow information in Britain, Chisholm and O'Sullivan (1973, p. 32) noted that:

> The data were obtained from a survey of the operations of the road haulage industry. Thus, the origin and destination of goods means only the origin and destination for the particular movement: the same goods may appear again as a separate movement. For example, the goods moving into a wholesale warehouse will have that building as their destination; when a delivery is made thence, the warehouse will be recorded as the origin. Consequently, the freight flow data, while accurately representing the work done by the transport sector, do not faithfully reproduce the flows from first origin to final destination.

That this really is a significant problem has been forcefully demonstrated by Gordon (1973). He compared the road freight data referred to in the above passage with data from the census of production. For this purpose, seven commodity groups were identified. With the exception of coal and coke, much of which moves by rail and also coastwise, the road haulage industry apparently moved between 117 and 291 per cent of the tonnage produced of the various goods. In other words, the double counting ranged from 17 to 191 per cent. To overcome these data deficiencies requires a massive amount of specially gathered information, something that is very expensive to do and requires much effort on the part of respondent firms as well as the researcher.

The second problem is illustrated in Figure 24, which shows twelve locations enclosed within two circles. Two of these locations, A and B, have connections with all the other locations, these connections being depicted by the straight lines. If the geographical unit used for recording purposes is the larger circle, then for both A and B all the links are intra-zonal. If the smaller circle is used, A still has the majority of its links within the area (seven out of a total of eleven), but for B seven of the links now cross the boundary of the recording area to destinations within the smaller circle. It is immediately apparent that the size and shape of reporting areas will determine

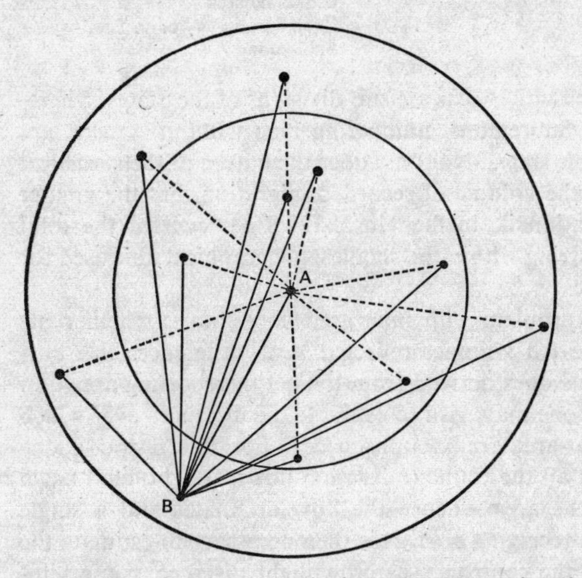

24. The effect of varying the mesh of reporting units on the volume of recorded transactions.

the number of movements that will be recorded. Because very many observations are required to establish the geographical pattern of flows, it is almost always essential to aggregate the punctiform origins and destinations into areas of greater or lesser extent, with the result that there is no escaping from the difficulty we have noted.

Now the shape and size of the geographical units used for statistical purposes vary from one region to another and may also vary over time. This fact poses two problems, one concerning the measurement of movements between areas and the other intra-zonal movements. The former issue is vividly illustrated by migration data for the United States (Schwind, 1971, p. 4). Of the 159 million inhabitants aged five years or older recorded in the 1960 census, the proportions that had changed residence in the previous five years were:

Total who had changed residence:	47·3%
of whom:	17·5% crossed a county boundary
	14·3% crossed a boundary of a State Economic area
	11·7% crossed a boundary of an Economic Subregion

The 509 State Economic Areas are sub-divisions of the States, whereas the Economic Subregions, numbering nine without Alaska, are aggregates of whole states. Not only does the nature of the statistical areas used affect the volume of recorded migration, but the greater part of the migration is, in fact, local (29·8 per cent of the total population), occurring within the smallest geographical unit used for statistical purposes.

Thus, the first problem with inter-area flow data is whether the variations in reported volumes reflected 'real' differences in transactions or specious ones derived from the net of reporting areas. In the second place, one may wish to estimate the distances over which the recorded inter-area transactions occur. For this purpose, one must assume that all the traffic (migration flows, freight, etc.) has a single origin in the area whence the flow originates and a single destination in the receiving area. It is then necessary to estimate the distance between the centroids, as crow-flight distance, route kilometres, time or cost. This estimating procedure is clearly open to numerous hazards, but these are much less than those involved in estimating the average distance for transactions within a reporting area.

If one knows the volume of transactions occurring within an area, how can the mean distance be estimated? One method was resorted to by Chisholm and O'Sullivan (1973). If in Figure 24 the origins and destinations are uniformly distributed, the average distance between them will be two-fifths of the square root of the area. This estimate involves heroic assumptions. Better approximations can be obtained but the techniques for doing so have only recently been developed and are laborious to apply (Taylor, 1971). The fundamental principle is to take each statistical area and lay over it a fine rectangular lattice; in this way, allowance can be made for its irregular shape and an estimate made of the distribution of intra-zonal distances.

Altogether, therefore, there are considerable problems involved in making an accurate description of movement patterns. However, once the hurdles of estimation have been surmounted, the data are in a very convenient form as an origin–destination matrix. This means that efficient procedures are available to obtain the maximum amount of information from the figures, as we shall see in the following section.

MATRIX METHODS: REGIONALIZATION OF FLOWS, CHANGES OVER TIME

As Rogers (1971, p. 1) notes, mathematicians invent new kinds of numbers when it is useful to do so, that is, when there is a problem that can be more conveniently handled in a new manner. Matrices are 'one of the most recent and most successful new kinds of numbers', and have come into widespread use only since the last world war. As we have already seen (Figure 10, p. 59), matrix notation provides a convenient transformation of a network so that the data may be readily manipulated. We have also seen that point patterns and other kinds of data can be examined very conveniently by conversion into lattices, which in fact can be readily presented in tabular or matrix format. In the present context, it is not necessary to dwell on matrix algebra techniques since they are very adequately described in numerous texts (for example, Mills, 1969; Rogers, 1971). The matter of immediate moment is to identify some of the underlying principles, first for the regionalization of flow data and second for the analysis of changes occurring over time.

In Chapter 3, reference was made to the grouping of areas to form regions, using an iterative procedure by which the 'distance' separating regions is calculated and the closest pair amalgamated. A somewhat similar procedure can be used for identifying nodal regions on the basis of flow data, for freight, passengers, telephone calls, etc. Figure 17(a) (p. 77) is a symmetrical matrix, such that the distance between A and B is identical to the distance between B and A. When dealing with flow data, this symmetry is lost. Thus, for example, Figure 25 is also a 4 × 4 matrix and contains precisely the same summation of values as Figure 17(a), but the individual entries differ somewhat and are no longer symmetrically arranged. Furthermore, the figures now represent flows, of goods, people, etc., not distances as in Figure 17. Area A has the largest volume of receipts of all the regions (forty-two units) and is therefore treated as the 'end point'

destinations

origins	A	B	C	D	Σ
A		(12)	9	8	29
B	(20)		6	6	32
C	10	3		(20)	33
D	12	9	(15)		36
Σ	42	24	30	34	130

25. A hypothetical origin–destination matrix of transactions.

of the system. For all four regions, the largest outflow, that is, the largest entry in the row, is circled. As region A has already been identified as the end point of the system, attention turns to region B. Its largest outflow is to A, and therefore B is treated as part of A's nodal region. However, for C the largest outflow is to D, and *vice versa*. As D is larger than C, D is treated as the centre of a second nodal region which includes C.

The identification of nodal regions in the above manner is based on the ideas developed by Nystuen and Dacey in 1961. Berry (1966) applied these ideas to India, using data on commodity flows between regions, a matrix of 36 × 36 origins and destinations. Four functional regions emerged, focusing on Delhi, Bombay, Calcutta and Madras (see Figure 26). At a very different scale, D. Clark (1973) examined the spatial structure of telephone calls in Wales to show the extreme dominance of Cardiff in that country's communications system.

Berry visualized that dynamic situations could be described as a succession of data matrices (Figure 4, p. 28) arranged sequentially in time. Between two time-periods, the information recorded in any one cell might or might not change. Given two matrices, one representing time t and the other $t + 1$, we may envisage a third matrix that describes the change-of-state. Thus, a most advantageous property of the matrix format is that it can accommodate statements of change

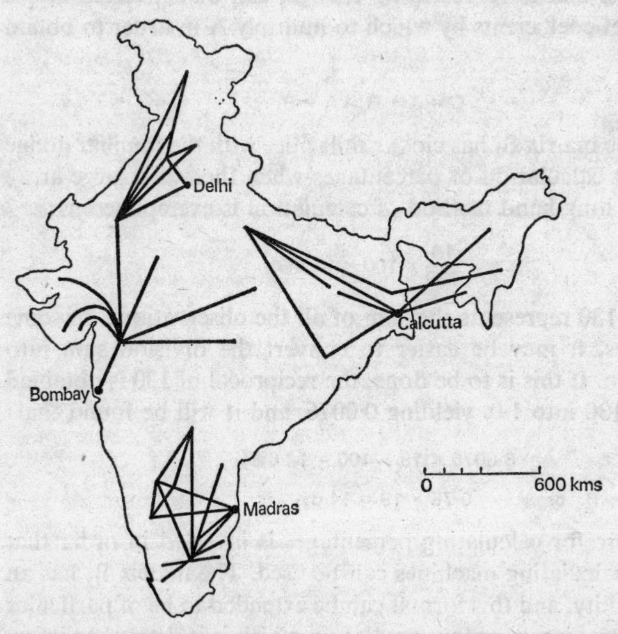

26. India: functional regions based upon total freight volume moving into trade blocks. (*Source: Berry, 1966, p. 156*)

just as readily as statements concerning situations at a moment of time. Furthermore, if we start with the three matrices mentioned above, with A and C representing the initial and terminal states and B the intervening changes, then

$$C = A + B$$

It follows that if one has data for only two (any two) of the above matrices, the third can be derived very easily, by calculations that can be done with quite simple equipment if necessary. By comparison, if the same data are represented in map form, it is relatively speaking a very laborious task to transform any pair of the maps into the third; or, alternatively, recourse must be had to quite sophisticated and expensive machinery.

Two developments from this basic idea are easily recognized. Matrix B, the matrix representing change, can be converted into a matrix (B_1) of coefficients by which to multiply A in order to obtain C. Thus,

$$C = A + B = A \times B_1$$

The use of the matrix B_1 has close similarities with the familiar dodge for the quick calculation of percentages when there is a large array of data. The long-hand method of calculation is exemplified as:

$$\frac{19}{130} \times 100 = 14{\cdot}6\%$$

In this case, 130 represents the sum of all the observations. In some circumstances, it may be easier to convert the division sum into multiplication. If this is to be done, the reciprocal of 130 is obtained by dividing 130 into 1·0, yielding 0·0076, and it will be found that:

$$0{\cdot}0076 \times 19 \times 100 = 14{\cdot}6\%$$

or

$$0{\cdot}76 \times 19 = 14{\cdot}6\%$$

The procedure for calculating percentages is inverted in order that mechanical calculating machines can be used. The matrix B_1 has an analogous utility, and this format can be extended to be of particular use in industrial input–output studies, in which it is desired to know the activity changes of all industries for a given change in final consumption. In the second place, we can conceive of the change from state A to state C as mediated by a set of transition probabilities, that is, the likelihood of given changes occurring. An early geographical application of this idea was Morrill's (1965) simulation of the evolution of settlement and communication patterns in southern Sweden on the basis of assigned probabilities of particular investments occurring at specified places. Hägerstrand's simulation of innovation diffusion was even earlier (see also p. 162).

SPACE–TIME BUDGETS

One of the more exciting developments in recent geographical work is the elaboration, mainly by Swedish workers, of the concept of space–time budgets. Pred (1973) has reviewed the development of 'time-geography' as a problem firstly of accurate description and secondly of fitting suitable models to the patterns identified. In his words:

> Throughout Hägerstrand's writings on migration and diffusion there is a common underlying theme linking social communication networks and *the changes in time and space jointly experienced by the individual and society as a whole.* In recent years, Hägerstrand has experienced a growing preoccupation with 'the fate of the individual human being in an increasingly complicated environment'. This combination of circumstances has led him and his research associates to embark upon an exciting and ambitious effort to devise a 'time-geography' model of society for the purposes of guiding urban and regional planning and locational policies in general.
>
> Every individual or household is surrounded by an *environmental structure*, or a pattern of resource and activity alternatives (water, food, other goods, job opportunities, services, information, social contacts, leisure-time possibilities, etc.) that are necessary to satisfy needs and wants and which are unevenly distributed in time and space. The environmental structure is relative to the individual; its composition depends on his information and economic resources and his psychological make-up. The time–space movements of any individual confronted by a succession of environmental structures can be depicted graphically by compressing space to a two-dimensional surface and representing time along a vertical axis. Thus, in earning a living and filling his informational, social and recreational needs and wants each individual wanders over an *individual path* which commences at a *birth point* and terminates at a *death point*. Depending on the analytic perspective desired, the individual path may be strictly defined either as a *daily path* or a *life path* through the use of time- and space-coordinates. While participating in production, consumption and social activities the individual stops at physically permanent *stations*, or areas of unspecified diameter where movement is not observed over time. The station concept is quite flexible in terms of both its space and time scales. For instance, 'what from a life-path perspective can be

regarded as a station – for example, a city of residence – is dissolved into a group of stations from a daily-path perspective'.

(Pred, 1973, pp. 36–8)

This approach by Hägerstrand and his associates is a logical derivative from the idea which we have already explored in Chapter 2, namely, the idea of mapping space in terms of time rather than geographical co-ordinates. The novelty is to treat human actions as occurring in a three-dimensional space, one vector of which is time, and to classify activities according to the kind of station occupied, or transfers between stations. As Pred shows, these ideas have been put to use in a variety of Swedish studies, demonstrating the practical utility of this descriptive method. One of the problems amenable to analysis in these terms is the layout of a new university on a campus site, where the space relationships of buildings are closely affected by the feasible space–time budgets of both students and staff.

REGULARITIES ABSTRACTED FROM TWO-DIMENSIONAL SPACE

We have already noted the astonishing regularity of the city size and city rank relationship, and the analogous regularity with which population density declines with distance from the city centre (pp. 85–8). Passing reference was also made to the systematic decline in interaction between places as the separating distance increases, the so-called distance-decay function. In the present context, the main points to note about distance-decay functions are that they are by far the most common of all the regularities abstracted from two-dimensional space, that they have been identified at all geographical scales from intra-urban shopping habits to international trade transactions, and seem to operate for all forms of spatial interaction. As they are given extended treatment in Chapter 5 in the context of gravity models, it would be repetitious to say much about them now, beyond noting that the geographer's distance-decay function is related to the economist's idea of price elasticity of demand. As distance from the place of origin increases, the cost of the transaction rises for the purchaser. This may happen in either of two ways. If the price is fixed at the point of sale, the buyer may have to pay for the transport costs of delivery. Alternatively, if, as in the case of retail shops, it is necessary to call in person, then there is an expenditure of time and probably also money necessarily incurred in making the transaction. Either way, the more distant the consumer is from the

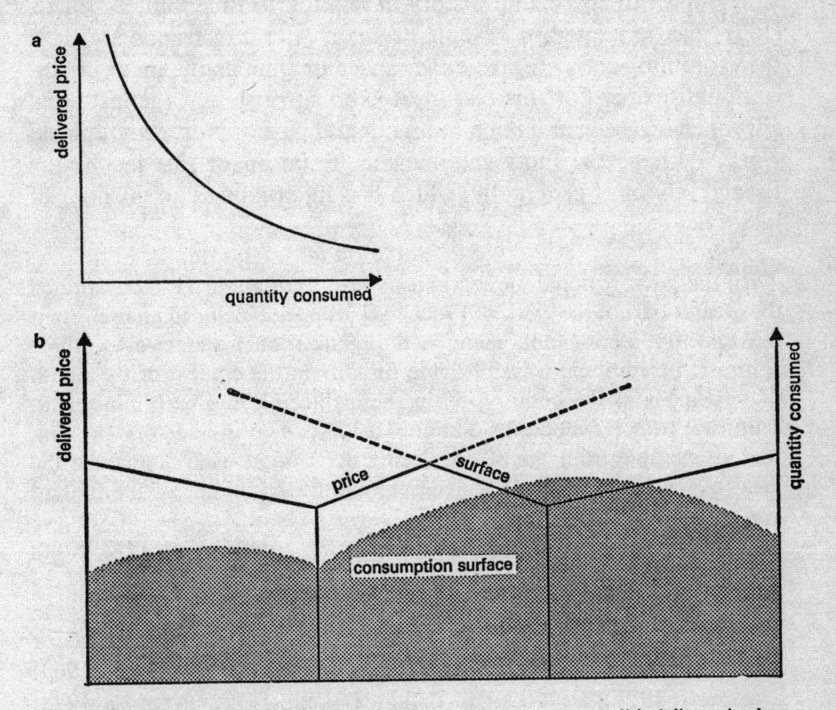

27. Prices and demand in space. (a) consumer's demand curve; (b) delivered prices in space with the related surface of consumption.

point of sale, the dearer the goods in real terms. As the cost to the consumer rises, so does consumption fall. Thus, the distance-decay function is a spatial demand curve, demand falling with distance; thus we have a major link between the empirical observations of spatial analysts and the theory of economists. Schematically, these spatial demand functions can be represented as in Figure 27: when rotated about the points of origin, they yield a demand surface.

THE HISTORICAL APPROACH

There has for long been much debate concerning the relationships between geography and history. In one view, geography integrates with respect to space, and history in relation to time, but, as Harris (1971) has reiterated, *all* events occur in *both* dimensions and it is therefore illogical to regard either space or time as the main organizational theme. For this reason, it is no surprise that the historical approach has been and remains an integral part of most geographical work. Perhaps the most extreme manifestation of this fact is the French school of geography, which had its origins as an offshoot of history.

> All the great regional monographs, which until about 1940 constituted the greater part of the work of French geographers, included an important retrospective section and many had organizational frameworks which contrasted traditional ways of living (usually in the context of the zenith of settlement in the countryside in the eighteenth and early nineteenth centuries) with the modern situation. History, as an explanatory element, was inseparable from the all-embracing description which remained the ideal of the French school of geography until the middle of the present century.
>
> (de Planhol, 1972, p. 30)

However, the tradition remains a powerful force among English-speaking geographers, as witness the publication of several major works since the late 1960s (C. T. Smith, 1967; Sylvester, 1969; Lambert, 1971; East et al., 1971; Baker, 1972; Pounds, 1973; Wrigley, 1961) as well as the appearance of successive volumes of *The Domesday Geography of England* under the general editorship of Darby (1952–).

In addition to the inescapable fact that all events occur in time, there are other reasons for the continuing vitality of the historical approach. Paradoxical as it may seem, the faster the rate of change and the greater the tempo at which institutions and knowledge become obsolete, the greater is the need for a knowledge of history, not the less. Although the past may become increasingly less relevant for an understanding of the present, it is equally true that an understanding of the present throws less and less light on both the past and the future – hence the need to study the past in its own right, for its own sake and as a guide to the future. Secondly, new data about the past are constantly coming to light and require re-assessment of previous work, a task rendered doubly necessary by the evolution of conceptual frameworks. Third and last, the very development of new concepts and theories generates a need for empirical testing, among other things to determine the degree of stability and universality in identified or postulated relationships; historical data are therefore in demand.

There are two traditional modes of historical study that have been employed by geographers. Both are alive and well today but are now being complemented by a newer quantitative approach. A logical corollary of the geographer's interest in landscape and regions (see p. 32) was the attempt to reconstruct the landscapes of the past. If these reconstructions of past geographies are arranged sequentially, one may visualize an unfolding 'film' of the area in question. The book that Darby edited in 1936 is in this idiom, as is Pounds' 1973 historical geography of Europe; indeed, the latter author introduces his book with the following words:

> This book was conceived more than twenty years ago as a series of pictures of Europe at a sequence of periods in European history, each to be linked with the one that follows by an historical narrative. It grew in size until first, it had to be divided into two parts . . . and then the linking narrative had to be omitted.
>
> (Pounds, 1973, p. xiii)

If the 'pictures' are close enough together in time, the illusion can be created of the continuity of development, just as a film creates from its succession of frames (still pictures) the appearance of movement (see pp. 92–3). In practice, the choice of dates for reconstructing the geographies of the past is strongly influenced by the availability of data, quite apart from the Herculean task that would be involved in creating a large number of closely spaced historical geographies.

Pounds' approach to the identification of patterns of evolutionary change contains two logical problems. Introducing his own historical geography of western Europe, C. T. Smith argued as follows:

> The reconstruction of past geographies is frequently regarded as the orthodox task of historical geography, and to some extent this concept has shaped the present study. But a series of static portrayals of geographical conditions in periods of stability is neither possible nor desirable in a work of this scale, even if such periods of stability could be found. It is the periods and processes of geographical change, of active settlement and colonization, of urban foundation and growth, or of industrial and commercial change, that stimulate most interest and that have been most significant in the formation of the landscape.
>
> (C. T. Smith, 1967, p. vi)

Note first Smith's scepticism regarding the notion of equilibrium conditions persisting for a sufficient length of time to be identified, a scepticism parallel to Kaldor's (1972) disenchantment with equilibrium ideas in economics. Second, Smith stresses the importance of processes generating change and thereby adverts to an age-old problem that we have already discussed, namely, how to infer the causal chain of processes from observed spatial patterns. The inherent difficulty of so doing encourages the scholar to examine as directly as possible the processes themselves, if not by direct observation then through contemporary records of one kind or another.

Thus, the second technique widely adopted by historical geographers has been the detailed examination of particular phenomena over time, compiling evidence concerning why and how events occurred. Interestingly enough, one of the pioneering works using this approach to historical geography appeared in 1940, thus coinciding in time with the appearance of work in physical geography also devoted to the study of processes; this was Darby's *The Draining of the Fens*, in which is reconstructed the history of canal construction, land drainage and agricultural changes in Fenland England.

Methodologically, historical geography has had close affinities with history and at first sight it may seem that historical geography is outside the mainstream of change in geography as a whole. To take such a view would overlook three considerations of the utmost importance. As noted in Chapter 3, there are various kinds of explanation that may be sought and it is far from evident that the only worthwhile kind is that in which explanation is equivalent to prediction. Thus, to improve our explanation of past events, though with no pretence of developing any capacity for post-diction thereof, remains an eminently worthwhile and beguiling matter. Who can fail to be intrigued by the identity of the 'dark lady' of Shakespeare's sonnets or by the reasons why Bristol's municipally owned docks missed the opportunity for major new investment on the south of the river Avon in the 1960s, and had to settle for a scheme to be completed in the mid-1970s that will probably be too little and too late?

One of the major continuing fascinations of historical geography is the way in which new concepts can be used to re-interpret the past. C. T. Smith refers to the application of the urban rank–size rule to England in 1377 in the following terms:

J. C. Russell has fitted a formula to the figures for English towns that he has derived from the 1377 poll tax returns:

$$n_r = C\frac{\left(1 + \frac{\sqrt{r-1}}{10}\right)}{r}$$

where r = rank of the city size, n_r = the population of the city of r rank in the region, and C = the size of the largest city. London, with 35,000, had about 1·5 per cent of the total population. York and Bristol, with 11,000 and 9,500 were both below the anticipated sizes of the second and third largest cities, which should have had about 20,000 and 13,000 respectively . . . Perhaps the larger towns were already relatively stunted by the centralization of administration and government as well as trade on London.

(C. T. Smith, 1967, pp. 304–5)

Russell's use of data for 1377 may be compared with the way in which Allen (1954), in one of the early investigations of the rank–size rule, sought to establish the generality of the relationship by examining data for several countries and time-periods, including Britain in 1086, shortly after the Norman conquest.

Another intriguing example of the possibilities opened up by re-examination of the past in the light of modern thinking is provided by Hodder's (1972a and b) study of Roman settlements in England. Figure 28 reproduces his map showing the distribution of urban centres:

> Thiesson polygons have been constructed around all the cantonal capitals and lesser walled towns in lowland Britain in the third and early fourth centuries A.D. Arcs of equal radius have also been drawn around some of the cantonal capitals to indicate the fairly even distance from them at which many of the lesser walled towns are located . . .
>
> . . . In view of the general correspondence of the arrangement of settlements . . . to geographical locational theories, it might be suggested that it was mainly behaviour relating to marketing and service provision which produced the pattern.
>
> (Hodder, 1972a, pp. 224–5)

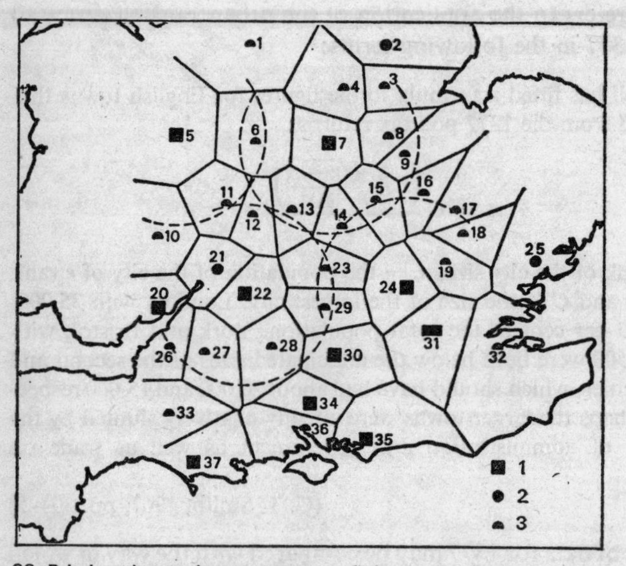

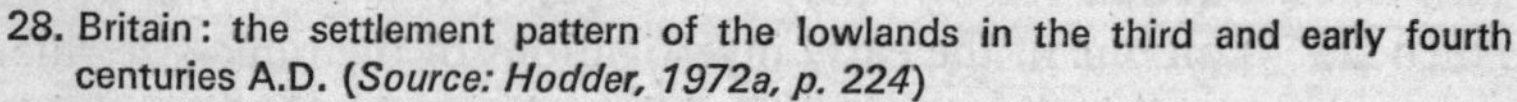

28. Britain: the settlement pattern of the lowlands in the third and early fourth centuries A.D. (*Source: Hodder, 1972a, p. 224*)

In addition to the process of re-interpretation, historical studies have an immensely important part to play in furthering the quantitative, model-building approach of modern geography. Very often, suitable contemporary data just are not available; alternatively, to establish the generality of a relationship, it is desirable to use historical as well as present-day information. History may therefore be used as today's laboratory. It was in this spirit that Hägerstrand (1953 and 1967) used a variety of data sets for his study of innovation diffusion, including information on grazing-improvement subsidies from 1928 and tuberculosis testing of cattle from 1899. Morrill (1965) went even further back into history in his simulation of migration and settlement development in a part of Sweden, 1810–20 being the earliest period for which he ran his model. Goheen (1970, p. 1) sought 'to explore the idea that the radical changes experienced by cities under the influence of nineteenth-century urbanization created a city comparable to the modern metropolis as conceived by social scientists'. He used Toronto, Ontario, as his case study and applied factor analysis, trend surface and other techniques to data for the period 1860 to 1900. A fourth example will be sufficient to establish the point beyond a peradventure. Pred's (1966) essays on the spatial processes of urban growth in the United States from 1800 yield a rich store of insight into the manner of economic and social change, blending theory with an imaginative use of empirical data; for example, he used the geographical distribution of registered patents as an indicator of the loci of innovation.

Some historical studies are specifically designed to throw light on contemporary problems, to indicate what should be done to alter trends that have been identified. One of the more significant publications in this idiom is that by Hall et al. (1973) on the processes of urban growth in Britain. A major feature of their work is the definition of 100 functional urban areas, using data on commuting as the major criterion. The functional areas so defined range in population upwards from about 66,000 in 1961. With boundaries established and held constant over time, the authors were able to analyse changes in population and employment over a span of several census years, and avoid the difficulties encountered through changes in the definition of the areal units used. Over the thirty-five years of the study-period, the proportion of the population living in the Standard Metropolitan Labour Areas of England and Wales has remained remarkably constant, at 77–78 per cent. However, whereas within these urban areas the population has been shifting from central to peripheral locations, the internal distribution of employment remained remarkably constant, at least until recently. As the authors point out, these differential trends have major implications for urban planning policies, affecting the provision of space for housing and improvements to the transport system, as well as many other features of the urban scene.

CONCLUSION

In a very real sense, this chapter occupies an intermediate position between the preceding and following ones. Chapter 3 is devoted to the ways in which static phenomena can be accurately and efficiently described. The present chapter continues this line of thinking but extends it to the examination of interaction patterns, in which new problems of data transformation emerge. Thus far, the geometrical tradition of the subject continues to be emphasized. At various points, however, it is convenient to look forward to the next chapter, to the explanation of the patterns observed. We have also made the transition from the analysis of form or structure to an interest in the processes that operate in the spatial domain to produce the observed patterns. In recent decades, geographers have become increasingly wary of inferring process from structure and have sought more direct evidence on the nature of the processes.

Of course, historical geography has for long been closely associated with the examination of processes operating in space and time. Historical geographers have painstakingly ransacked archives of all kinds in their search for evidence; York Minster held a key of major importance to unlock the secret of how the Broads were formed, as vast commercial peat diggings exporting the peat far and wide in England. But there has been a gap between our knowledge of events in the past and of processes operating in the contemporary world. Only relatively recently have geographers launched a serious campaign to bridge that gap.

For this purpose, the historical method which supposes that to arrange events in chronological order is to explain them is patently inadequate. Even if we assume that time has only one 'direction', from the past through the present to the future, it is necessary to look for links between events other than their relative location on the time dimension. Though the Romans founded a city on the site of what is now London, and indeed even gave that city its name, no one would seriously contend that the Romans had anything to do with the present size and importance of Britain's capital. All events occur in space as well as in time, and if there is a readily identified direction of causation in the time dimension – the more recent being affected by the less recent but not *vice versa* – the same cannot be said of space. And so it becomes necessary to seek out the frameworks within which to examine multi-directional systems of causation. This implies careful examination of the ways in which phenomena are linked in space, and especially the flows of goods, people and information between locations. Hence, in the earlier part of this chapter, our attention was turned to the analysis of movement patterns and the structure of networks – networks being the avenues for interaction.

However, to link together the study of forms and of the processes that give rise to them, accurate description is not enough. We need a framework of explanatory theory, to provide us with a rational understanding of the inter-linking of processes and their relationship to spatial forms. This is the subject of the next chapter.

5. Theories of Spatial Structure and Process

The theoretical underpinning of social
science is still very primitive.

(Shonfield, 1972, p. 430)

Explanatory theories may be regarded as the apex of the scientific pyramid, the goal to which we strive. However, this apex cannot exist without foundations. In the case of the geographical pyramid, these foundations consist first of the identification of spatial patterns and second of the examination of processes operating in the spatial domain. Broadly speaking, Chapter 3 relates to the former and Chapter 4 to the latter, while now in the present chapter we can turn to the question of theory. Although distinguished thus baldly for expository purposes, the stages of description and explanation cannot be separated so clearly in practice, since the recognition of a pattern depends on some prior concept of what is to be ordered and how – and hence some concept of explanation, however ill-formed. Nevertheless, as the body of observed facts expands and the corpus of recognized spatial patterns enlarges, the need for explanation – for theories and models – increases. It is therefore a matter of no surprise that geographers have become increasingly interested in theory; indeed, this is perhaps one of the main hallmarks of contemporary geography.

However, there seems to be a great deal of confusion over the meanings of terms and hence what it is that various workers are seeking to achieve. Consequently, a brief recapitulation of the main issues is necessary, if only to clarify for the reader the meaning of terms as used by the present author. For this purpose, it is instructive to take the 'scientific paradigm' in terms used by Hempel, namely:

$$\left.\begin{array}{l} C_1, C_2, \ldots C_k \\ L_1, L_2, \ldots L_r \end{array}\right\} \text{Explanans}$$

$$\overline{E} \quad \text{Explanandum sentence}$$

where $C_1, C_2, \ldots C_k$ are sentences that describe the initial conditions of the particular facts; $L_1, L_2 \ldots L_r$ represent the general laws on which the explanation is based; and E stands for the explanation derived from the two previous sets of sentences. An academically respectable discipline, many hold, is one that contains a substantial number of statements couched in the above terms, these statements being about non-trivial matters. The explanatory system described above is a general statement of the nature of a 'theory', and assumes that the laws have been fully established and verified. Thus, following the *Shorter Oxford English Dictionary*, a theory is:

> A scheme or system of ideas or statements held as an explanation or account of a group of facts or phenomena; a hypothesis that has been confirmed or established by observation or experiment, and is propounded or accepted as accounting for the known facts; a statement of what are held to be the general laws, principles, or causes of something known or observed.

This is the kind of theory that can be used to give precise answers to specific questions. Less powerful theory is that in which the general 'laws' ($L_1, L_2, \ldots L_r$) are not fully established, perhaps because quantitative values have not been generally agreed but only the direction of the causal relationships.

The word 'model' has become a grotesquely over-used term, both in geographical circles and more generally. As a result, there is confusion over its meaning, just as there is with the word 'hypothesis' (Newman, 1973). As used by Haggett (1965) and others, the term 'model' seems to subsume any representation of reality, whether this consists of data transformations (a map, for example), a theoretical construct or a simplified working system that replicates some features of the real world. The *Shorter Oxford English Dictionary*, however, clearly indicates that a model is something concrete and tangible, in the form of a drawing, scale replica or something similar.

Now, to be of any use, a theory must be conveyed to other people and must therefore be mapped into a language – literary, numerical or symbolic – and thereby converted into a model. In this sense, model and theory may properly be regarded, if not as synonymous, then as two sides of the same coin. But the term model also carries the connotation of a replica constructed without there necessarily being any real knowledge of how the actual object or system works, that is, when there is no theory: Michelangelo made a splendid model (statue) of David without needing a *fully* articulated theory of how the human body and brain function. The relationship between theories and models is therefore not symmetrical; whereas all theories have a corresponding model, not all models are provided with an appropriate theory. To avoid confusion, therefore, use of the term 'model' ought to be restricted to tangible representations of reality, irrespective of whether the representation is or is not based on an articulated theory. Theory should be restricted to the articulated systems of ideas, schematically represented on p. 122. In this sense, theory may be fully specified, with all the laws established, and therefore able to yield precise predictions. Alternatively, the 'laws' may be hypothesized, in which case the explanations offered are conditional and subject to verification. In this second case, the verification may be by reference to the real world, in which case the theory is positive in character, or if the purpose is normative, then verification consists of checking the internal consistency of the theory and the initial postulates, but not a comparison of the end result with the actual situation in the world. This is a point to which we must now turn.

Now a number of geographers, among whom Chorley, Haggett and Harvey have been notable, are of the opinion that geography's next major task is to develop geographical theory and laws. Harvey has certainly expressed himself as sanguine that this can indeed be successfully accomplished.

> There can be no doubt that the natural sciences (and physical geography) are in an advantageous position relative to the social sciences (and human geography), since the laws they possess are, by and large, better substantiated (and consequently require less in the way of assumptions in order to be employed) than are the laws in social science (and human geography). But . . . it cannot be inferred from this that powerful laws cannot ever be developed in social science and human geography. There is every reason to expect scientific laws to be formulated in all areas of geographic research, and there is absolutely no justification for the view that laws

cannot be developed in human geography because of the complexity and waywardness of the subject matter.

(Harvey, 1969, p. 113)

At this point, it is useful to examine a widespread misunderstanding (see, for example, Guelke, 1971). Guelke avers that he cannot conceive of geographical theories and laws that will meet the exacting specifications set out by Hempel and endorsed by Harvey. In particular, he is afraid of the multiplication of *untestable* theories, citing central-place theory (p. 143) as a case in point among others; the fear is that energy will be directed mainly to accounting for the discrepancies between fact and theory. Such a view incorporates a double misapprehension.

The first of these can readily be recognized from Figure 29. The

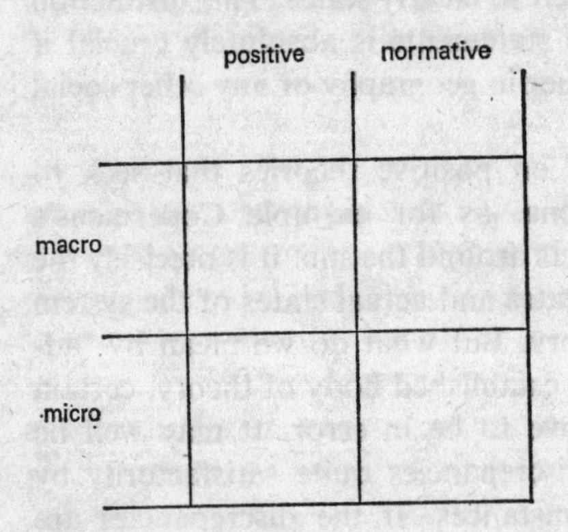

29. Kinds of theory.

essential characteristic of central-place theory, as developed by Christaller and Lösch, is its normative character, that is, the fact that the theoretical construct is not intended to show how the world actually is organized, but how it should be. In the words of Lösch:

Theory may be compared with reality for various ends, according to the sort of theory held. If it is to *explain* what actually is, the examination attempts to discover whether it started with a correct idea of its subject and arrived at an explanation that not only seemed possible but also corresponded with reality. On the other hand, if theory is to *construct* what is rational, its assumptions may still be tested by facts, but not its results.

(Lösch, 1954, p. 363)

Thus to criticize normative theory because it fails to yield positive results is to tilt at windmills. However, it is easy to see how this mistake can be made when it is remembered that the formal 'scientific paradigm' as set out above and associated with Hempel is only valid in a positive context, for explaining how phenomena in the real world actually occur and behave. To use this framework of thought to determine how the real world should behave, it is necessary to make *assumptions* about both the sentences that describe the initial conditions (C_1, C_2, . . . C_k) and the laws on which the 'explanation' rests. This kind of theory is not amenable to verification in the normal scientific manner, as Lösch so clearly states. This distinction between normative and positive statements is absolutely crucial if confusion is to be avoided, whether in geography or any other social science.

If, then, attention is focused on positive theories that seek to account for observed phenomena, as for example Copernicus's theory of the motion of the planets around the sun, it is precisely the discrepancy between predicted states and actual states of the system that stimulates advances in theory. But what do we mean by 'advance' in this context? From an established body of theory, certain predictions are made which prove to be in error. It may well be possible to account for these discrepancies quite satisfactorily by invoking special rules or circumstances. If the discrepancies are numerous, the special cases multiply and either the theory becomes increasingly complex and cumbersome, or inconsistencies are introduced, or both. A theoretical advance is said to occur when a modification of the original theory happens or a new theory is propounded, which simultaneously achieves three things:

1. Accounts for the previously noted discrepancies.
2. Maintains or enhances the accuracy of the predictions made.
3. Achieves parsimony in assumptions and formulation.

Kepler's substitution of elliptical for circular orbits of planetary motion represented just such a theoretical advance, fulfilling all three conditions. In the geographical domain, equivalent advances were made in the nineteenth century through the full realization of the erosive power of ice and in the present century by acceptance of the idea of continental drift (under the modern guise of plate tectonics). Of equal importance in human geography was Christaller's formulation of the reasons for a hierarchical ordering of cities (see p. 143) and Wilson's (1970) examination of entropy concepts underlying gravity models of spatial interaction.

In the two previous chapters, attention was focused on the identification of spatial patterns, which implies an inductive approach to the formulation of theory. By definition, inductively derived theory must be positive in character, relating to the real world. Very much less effort has been directed by geographers to developing deductive theories, whether normative or positive. The balance has shifted somewhat in recent years, at least with the widespread acceptance of normative theories within geography, albeit mainly derived from other disciplines. Perhaps the most important next step is a general realization that a healthy academic enterprise needs *both* inductive and deductive modes of thought, perhaps in roughly equal measure, for in practice the one reinforces the other.

While there is a need for both positive and normative theories, another distinction must be clearly made. Figure 29 distinguishes between macro- and micro-theory, that is, distinguishes between theories that apply to the whole population being considered, for example all the cities of the Löschian landscape or all intra-urban migrants, and theories that apply to individual members of such populations; the former is termed 'macro' and the latter 'micro'. The essential point to bear in mind is that the assumptions suitable for one scale of analysis are most unlikely to be appropriate at the other. Yet, as Chisholm (1971a) has shown in the case of location theories, micro-analysis assumptions have in fact been carried over into macro-studies and with rather unfortunate results.

Following from the above discussion is a general point which the present author believes to be of fundamental importance. It is widely accepted that if one changes the geographical scale of an inquiry, or the time scale used, the mode of analysis may have to be changed and that conclusions valid at one scale may not necessarily apply at another (for instance Haggett, 1964). An exactly analogous proposition applies concerning shifts between macro- and micro-theory, and theory that is normative or positive. Consequently, one must visualize families of theories, with each theory tailored to specified conditions. It would be nice to think that each theory could be linked to others so that the whole map of knowledge could be articulated by an integrated set of theories, but this ideal situation seems to be unattainable. Thus, one must accept the existence of separate theories which may be only very loosely related to each other.

There is one last issue to which a few words must be devoted. As we have already seen, geography has traditionally been rich in facts but poor in theory, though that is a situation which fortunately is changing. But whence come the theories that we now use? Many geographers are anxious because the amount of 'indigenous' theory developed within the subject is small compared with the amount borrowed from other subjects (see, for example, King, 1969b; Sack, 1972). This is an anxiety that I do not share. No one thinks the worse of a meteorologist because his basic theory is derived from physics, nor of an economist who makes behavioural postulates taken from the work of psychologists. It is not the provenance of theory that counts but the use to which it is put. Since geographers have indeed made good use of the available theories, and also contributed to the extension of theory, there is little reason to worry about the identity of the subject as expressed by a corpus of 'geographical' theory. In this respect the position is entirely analogous to the problem of defining a subject in terms of the 'facts' it studies; as we have seen in Chapter 2, particular facts are not the exclusive property of one subject, and the same proposition applies to most, if not all, theories.

In the pages that follow, only a selection of theories will be presented. The purpose is not to provide an encyclopedia of theoretical advances but to show some of the main highways along which geographers are travelling. Some roads lead to the same destination, while others take us to quite other places. However, these main roads – if we may pursue the analogy somewhat further – are linked in varying degree. Though presenting an incomplete map of geographical work, they do mark out the intellectual territory with which geographers are familiar. Of necessity, we shall pass by innumerable minor roads that lead to intriguing problems, but we do not have the time to explore even a fraction of them. For the purpose of clarity, theories have been divided into two groups, normative and positive. Within these broad headings, it is not possible to maintain a rigid distinction between micro- and macro-theories, since some are suitable for use at various scales. Thus, while the boundary between positive and normative theory is fairly clearly-cut, the same is not true between scales of analysis.

NORMATIVE THEORY

At first sight, much normative theory is presented as being mainly relevant for the analysis of regional or national spatial patterns. An excellent example is provided by von Thünen's famous zones of agricultural production encircling a central city, to which all farm produce is sent (Chisholm, 1962; Hall, 1966). Nevertheless, the basis from which this theory of regional allocation is derived is essentially that of the behaviour of individual farmers. The concentric pattern of agricultural production arises because neighbouring farmers are assumed to arrive at the same decision regarding their production pattern, both as regards type of crop (or livestock), and the manner of production.

For the moment, let us concentrate on a second feature of normative theory. To specify what the location pattern should be, it is necessary to establish some criterion and formulate the problem in such a way that a 'best' distribution can be found. As a farmer, the criterion by which we might choose our production schedule would be profits; the best farming system is that which, for the given location, maximizes these profits. We may conceive of profits as an objective function, which in this case is to be maximized. In other cases, the aim may be to minimize the objective function; for example, it might be desired to minimize the costs of production, rather than explicitly maximize profits.

If we emphasize the nature of the objective function, we have a convenient way in which to classify normative theories. Other approaches might emphasize the historical evolution of normative theories, or classify theories by the kind of phenomena involved – agriculture, manufacturing industry, towns, etc. Both of these approaches would involve a significant element of repetition and have therefore been rejected; the nature of the objective function provides a more convenient and more general framework within which to work.

Production costs

Although he was preceded by several German scholars, it is primarily to Alfred Weber (1909 and 1929) that English-speaking scholars turn for the formulation of basic industrial location theory. He visualized the problem facing a firm that wishes to locate a new plant. The sources of material inputs are assumed as known, as well as the markets to be supplied. Further, the scale of output has been decided and it is assumed that the relative quantities of all inputs are known (that is, the production-function is taken as given). Unit transport costs are identical at all locations on the uniform plain. The initial problem then resolves itself into finding that location which minimizes total transport costs, which Weber assumed was equivalent to minimizing the tonne-kilometres of freight.

In Weber's original formulation, the factory could potentially be located at any point on the plain. Therefore, with an infinite number of possible locations, there were considerable computational problems in finding the optimum site. These problems can be greatly reduced if the feasible sites are reduced to a finite number, since it now becomes possible to arrange the information in matrix format to show the transport costs involved at each location. A relatively simple search procedure will then allow one to identify the site with the lowest aggregate cost of transport.

Weber and his followers extended the argument by noting the spatial variation in production costs other than those relating directly to the transport costs of assembly and distribution. What Weber termed the attraction of labour supplies, others have generalized to include all other production costs and particularly those which vary on account of agglomeration economies. These latter are scale economies external to the firm. Thus, if several firms locate together they can share services like sewerage and insurance facilities, the unit costs of which will therefore be lower than if they had to be provided for scattered plants (Hoover, 1948; Isard, 1956; Estall and Buchanan, 1961 and 1966).

The Weberian approach assumes that each firm is the decision-making unit and that decisions will be taken in the light of full knowledge of all the relevant information. Spatial order, or pattern, will emerge as the result of these individual location choices. More important, though, is the question whether cost minimization is satisfactory as the criterion for a location choice. In the case of a manufacturing company this assumption will only be valid if it can be assumed that revenues are unaffected by the location choice, because only under this assumption will the point of minimum cost equal the point of maximum profit. The assumption of invariant revenue is patently not tenable and therefore Weber's approach must be regarded as too simple, at least in the general case. However, some problems can be very usefully set up in cost-minimizing terms, as we shall see in the next two sections.

Linear programming

The basic concept of linear programming is really quite simple and has been described in a number of texts (for example, Yeates, 1968). As a technique, it is designed very precisely for the kind of problem we have been discussing, that is, how to maximize or minimize some clearly specified function subject to stated constraints. Yeates imagines the problem facing a land-reclamation agency which wishes to minimize the costs of reclamation subject to the requirement that at least 10,000 hectares be reclaimed in total and that within this total there must be at least 4,000 hectares devoted to urban uses and 5,000 to agriculture. Reclamation costs are known, £400 a hectare for urban land and £300 for agriculture. The answer is intuitively obvious, and is:

$$(4{,}000 \times £400) + (6{,}000 \times £300) = £3{,}400{,}000$$

Underlying the intuitive answer is a logical structure, of the following form: what is the total reclamation cost for each of the feasible combinations of land quantities and of these which is the combination that will satisfy the constraints at the lowest cost?

In the trivial case here presented, it is possible to go straight to the answer. For more complex situations, it is necessary to set up a search procedure (an algorithm) whereby an initial *feasible* solution that satisfies the constraints is obtained and then by successive adjustments the solution is improved. In this way, the answer converges on the *optimum* arrangement.

Linear programming has been put to use in two major study areas relevant to geographers. For the individual firm, whether a farm or factory, it is a very helpful technique for deciding the optimal combinations of inputs and outputs. It is also a powerful tool for examining the efficiency of freight (or other) movements over a network. An early application was by Land (1957), who compared Britain's actual pattern of coking-coal freight-flows by rail with the optimum as determined by linear programming. For the purpose, he used 154 sources of supply, represented by coal pits, and 65 destinations, the locations of coke ovens. Assuming that the actual output and consumption at the respective points was 'correct', the problem was to find that allocation and routing which would minimize the total tonne-kilometres of freight movement. Somewhat surprisingly, this most efficient pattern of distribution represented only a 10 per cent saving over the actual pattern of flows.

More recent work by Chisholm and O'Sullivan (1973) was based on road freight divided into eleven rather crude commodity groups. They found that for the seventy-eight-zone system covering Great Britain the correspondence between actual and expected flows was surprisingly good. With the exception of two commodities, the correspondence was always over 50 per cent and exceeded 80 per cent in five cases. Taking just the twenty-four major cities of England and Wales, the level of explanation exceeded 90 per cent in all cases except the miscellaneous commodity group 'other manufactures', for which r^2 was only 0·49. The inference seems fairly clear that with the existing distribution of population, and hence of supplies and demands, the macro-flows of freight in Britain are remarkably 'efficient'.

Although linear programming has been presented as a cost-minimizing technique, the reader will readily appreciate that it is equally suitable for finding maximum-profit solutions – *if* the relevant constraints can be specified and the requisite data are available.

The partitioning of space

Innumerable problems present themselves concerning how most efficiently to divide a territory into administrative/functional units of one kind or another. There are two basic cases to consider:

1. If the administrative centres are defined, what is the most efficient pattern of regional boundaries? The inverse of this problem is to locate the centres if the boundaries are taken as given.
2. Alternatively, it may be desired to locate the centres and the boundaries simultaneously.

Both cases can be tackled by techniques that are explicitly linear programming in nature or have a close family resemblance to it.

Yeates (1963) conducted an interesting experiment on school catchment areas in Grant County, Wisconsin. There were thirteen high schools, the locations of which he accepted as given, as also their sizes. Thus, the problem was to allocate all the pupils to the nearest available school, subject to the constraint that each school must have the precise number of pupils it already catered for. The result would give the minimum aggregate cost of transport for the school population. To obtain the answer, Yeates divided his study area into mile-square sections, that is, a regular lattice, and used these areas, with their resident population of pupils, as the units for assignment. Comparing the results of his linear programming exercise with the actual catchment-area boundaries, he concluded that 18 per cent of the cells were incorrectly assigned (Figure 30). The inefficiency of the system was actually less than this figure, since the cells that should be re-allocated were relatively sparsely populated.

The problem to be solved could easily be inverted. If the catchment boundaries were defined, the size of each school would automatically

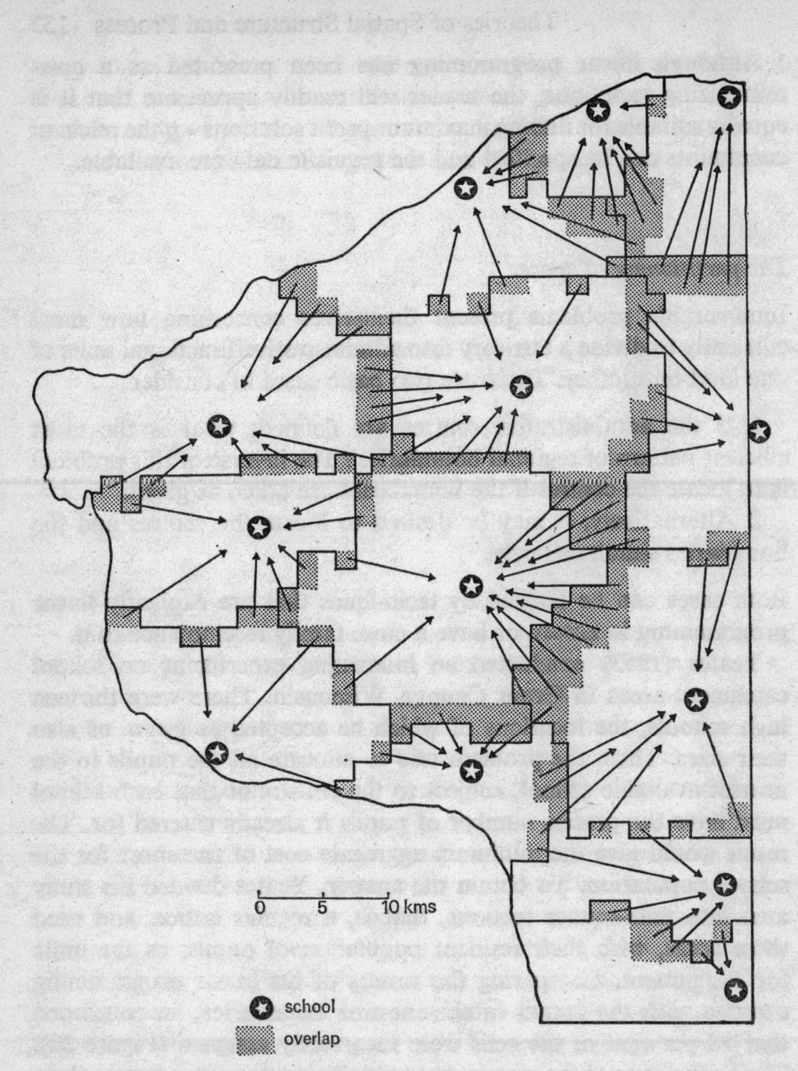

30. Grant County, Wisconsin: the overlap between the actual set of school-catchment boundaries and the spatially efficient set of boundaries. (*Source: Yeates, 1968, p. 112*)

be determined. Hence, the problem would be to find the location for each school within its catchment area that minimized the aggregate travel distance for its pupils.

Haggett and Chorley (1969) point out that one of the more important problems to which optimizing procedures of the above kind can be applied is the definition of electoral areas. In fact, a variety of objective functions can be devised to achieve different aims, from minimizing travel costs to maximizing the electoral advantages of a particular political party. Altogether, there are now over twenty procedures (algorithms) for electoral districting (Taylor, 1974).

A problem more difficult than that investigated by Yeates arises when the centres and the regional boundaries are to be located simultaneously. Törnqvist et al. (1971) developed a technique to meet this situation and Hirst (1973) applied it to Uganda, where the government had decided to establish nine provincial centres to administer the country. Hirst laid a rectangular lattice over the country and obtained the population for each cell. Initially, he took as given the locations of the nine provincial centres selected by the government. Each cell of the lattice was then allocated to the nearest centre. It was thus possible to calculate the 'transport costs' of this system. The population of each cell was multiplied by its distance from the allotted centre and the resulting products summed for the entire country. The next step was to ask: Can a better distribution of the nine centres be found that would reduce the 'transport costs' of the whole system?

To answer this question, Hirst took as his starting point the solution described in the previous paragraph. Then follows an iterative procedure in which the provincial centres are allowed to 'move' from one cell to another. At each step, the total transport cost for the whole system is calculated. If the 'movement' of a centre results in a lower summation of costs than was obtained at the previous step, it is regarded as an improvement. The process is repeated until there is no move that will result in an overall saving. Hirst's results are shown in Figure 31; compared with the nine centres selected by the government (31a), his system of nine provincial capitals and associated administrative areas (31b) would yield a saving in transport costs of 22 per cent, or almost one-quarter.

Much of the pioneering work on the choice of locations for facilities such as gymnasia, schools and hospitals, and on the delimitation of urban spheres of influence, was undertaken by Swedish geographers. From the early 1950s onwards, their studies attracted a

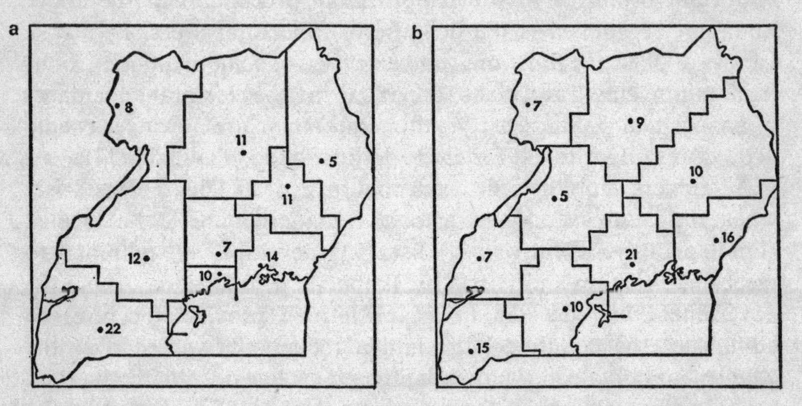

31. Uganda: administrative divisions of the country. (a) The optimum allocation of the population grid to the nine proposed provincial headquarters. The numbers represent the percentages of the total population which should be assigned to the respective centres. (b) The optimal locations and capacities of nine facilities in Uganda to serve the total population. Distance exponent is 0·9.

great deal of attention in Swedish official circles, with the result that geographers were enlisted by the government to advise on a whole range of practical problems for which decisions were required (Pred, 1973).

Profit maximization

Whereas a cost minimization approach is clearly appropriate for a government agency or equivalent organization that provides a service for which there is no true market, for commercial organizations there is only a limited number of situations in which this approach is useful. As we have already seen, the Weberian approach to industrial location is inadequate because spatial variations in revenues are ignored. Explicit consideration must be given to profit maximization.

In terms of location theory, it is primarily to Lösch (1954) that we must turn for the relevant ideas.

As long ago as 1929, Hotelling pointed out that a firm may choose between two classes of strategy, or combine them in varying degree: strategies of location and strategies of pricing (or, more generally, of marketing policy). In the simplest case, the individual firm is assumed to take the external conditions as fixed and can therefore ignore the actions, or potential actions, of rivals. Even then, a firm must make numerous complex calculations. To the cost considerations previously discussed should be added the possibilities of substituting between transport costs and land costs (Isard, 1956). For example, many factories and offices have decentralized from London, New York and elsewhere because urban land prices are now so high that they more than off-set any accessibility advantages.

Even more difficult issues are faced when a firm attempts to assess the effects of its location choice upon its volume of sales. For a shop in an urban area, sales volume is materially affected by ease of access relative to the accessibility of rivals. This idea can be generalized in the following manner. As distance from the point of supply increases, so does the price to the purchaser, if by price we mean the commodity price plus 'cost' of getting it to its destination. As the delivered 'price' rises, so will consumption fall; demand is said to be price-elastic (see p. 113). Consequently, in reviewing its location options, a firm will need to estimate the delivered 'price' to consumers and their response to variations therein. In this way, location choices simultaneously affect both the revenue and the expenditure accounts.

If we imagine that several firms have made their own independent assessments of their most profitable locations, then it is inherently likely that two or more may compete for the same site. We may envisage this competition being resolved through the land market, on the basis that the highest payment for the preferred site will be made by the firm for which the *difference* in profitability between that site and its next best is the greatest (see Chisholm, 1966). In this way, land values represent the so-called transfer payments, or payments made to command the site and prevent it being occupied by other firms.

The idea of land values as the mechanism for allocating land uses derives from the earlier notion of 'economic rent', developed by von Thünen in 1826 (Chisholm, 1962; Hall, 1966) in the context of agricultural location patterns and, independently, by Ricardo. Economic rent is to be conceived as the *surplus* which a given volume of inputs of capital and labour can produce by virtue of being applied to land that is within the margin of use. Thus, the datum is the output from the marginal land in the system being considered, that is, the land which it is only just worthwhile using. Economic rent is the surplus above this datum, accruing to each parcel of land.

Von Thünen conceived of a uniform plain with a single central city, acting as the sole market and source of purchased inputs. The economic rent of each kind of land use would depend on location relative to the central city, and all farmers at a given distance would select that pattern of production which yielded the highest economic rent at that location. In this formulation, therefore, the objective function to be maximized for the whole farming system is the value of economic rent. This amounts to maximizing profits within the stated constraints, which in practice is approached by minimizing transport costs.

Though we may envisage the land market as the major mechanism for allocating land between firms and various land uses, this expectation is founded on a whole range of assumptions, which may only be valid when one firm is being considered, all else being taken as given. However, a fundamental difficulty arises in attempts to move from normative profit-maximizing propositions valid for a single firm seeking its optimum location/production/marketing position to useful statements concerning the spatial arrangement of a whole economic system. Chisholm (1971a) has reviewed efforts to solve the fundamental dilemma that is involved in making this transition from micro- to macro-analysis, a dilemma which can be stated quite simply. What are the conditions necessary whereby numerous firms competing for supplies, markets and locations (for their productive enterprises) can achieve a *stable* equilibrium situation? On the assumption that firms are seeking to maximize their profits, there appears to be no logically consistent set of assumptions available to yield a stable solution. The nature of this logical difficulty is well demonstrated by a passage from Lösch. He envisages a 'hidden hand' guiding the economy to equilibrium as the resultant of two forces, (i) the profit-maximizing propensities of firms and (ii) the maximum advantage for society as a whole, represented in this case by the maximum number of independent economic units.

> The point where these forces balance determines location. This equilibrium, born of the interdependence of locations, can be understood only through a system of general equations of location. As soon as the conditions expressed by these equations have been fulfilled the struggle for space dies down, and when the equations are solved the locations themselves are determined.
>
> (Lösch, 1954, p. 94)

The assumption made by Lösch is that all of the firms will simultaneously recognize the situation which is 'best' for the space-economy in question and will 'obey' the requirements that follow from it. This is hardly an acceptable proposition on which to build a normative theory of spatial organization.

Although it is clearly better to aim at profit-maximizing solutions wherever these are appropriate for normative theory, we have so far not progressed beyond the limited case of partial equilibrium. That is, we have been dealing with theory suitable for a single firm which can take the external environment as given. The difficulty is that such an approach does not permit one to say anything useful about the normative pattern for several firms in competition. In the succeeding sections, we will examine ideas that have been advanced to break out of this restrictive situation.

Game theory

Game theory is based on the idea of conflict and attempts to provide a framework in which to resolve the different interests of the participants as they each strive to obtain maximum benefits for themselves. The initial analogy is with a card game such as bridge, in which each player knows which cards he has in his hand and can therefore eliminate the possibility of other players holding them. Strategy is then based on assessments of how the cards are probably distributed among the other players, for which purpose the process of bidding provides a code of rules whereby inferences may be drawn.

To formulate this idea into 'operational' terms in the locational context, it is necessary to make some very stringent assumptions. If we conceive of two firms (players) faced with making a choice between two locations (strategies), we can formulate the problem as follows. For both of the firms there are two strategies available and we will suppose that both firms know what the benefits/losses of these strategies are. Furthermore, let us suppose initially that firm A

		player b	
		strategy 1	strategy 2
player a	strategy 1	6	5
	strategy 2	5	4

32. Game-theory pay-off matrix.

always wins and seeks to maximize his gains, whereas B always loses and tries to minimize his losses; they both have an objective function, the one the inverse of the other. The situation may then be represented as in Figure 32; the figures in the rows represent the gains to A, while the identical figures viewed as entries in the columns represent B's losses. It is assumed that each of the players, faced with the information in the matrix, makes his choice in ignorance of the choice being made by the other. However, inspection of the data immediately reveals that A will always choose strategy 1 – he will win at least 5 and might gain 6 units of 'profit'. For B, strategy 2 is to be preferred – he might lose as much as 5 but might get away with a loss of only 4.

This is, of course, a highly artificial situation but it does produce a clear-cut answer. However, even within the rigorously simplifying assumptions so far made, other figures in the pay-off matrix will yield an unstable situation – on the assumption, that is, that the game is played not just once but many times. If the game is amended to match more nearly the conditions of the real world, the calculation of the best strategy for both players very rapidly becomes extremely complex.

Isard and his colleagues have attempted to develop game theory in the context of industrial location choices, and their work (published over several years and collated as a book in 1969) has been reviewed by Massey (1968). Perhaps the most striking point to note is the lack of progress in putting real data into the pay-off matrix, at least in the field of industrial location. In the industrial and commercial context, the fundamental reason is that the information does *not* have an objective existence independent of the participants; they must each make their own assessment of the data, as they are not able to receive them from some disinterested third party. Thus, while game theory is clearly a useful way in which a firm can formulate its own approach to the task of finding an optimal strategy, and while the technique has undoubted educational/instructional value, it manifestly cannot supply a framework for identifying a stable and optimal spatial allocation for a whole system of competing firms.

However, if one of the 'players' is conceived to be 'nature', then a two-person game can be immensely useful. Of the seven examples cited by Abler, Adams and Gould (1971), six fall into this category – the seventh being a hypothetical case of two armies either side of a range of mountains considering their best strategies to control two vital passes. Where 'nature' is one of the players, then suitable observations of climatic or other variables will enable a reasonable estimate to be made of the frequency with which certain events occur. Thus, the human player can know with fair precision the frequency with which nature will 'play' specified strategies. Furthermore, experience will show what the pay-off to the human player is for the strategies 'played' by nature.

The African farmers of Jantilla, in the Middle Belt of Ghana, have grave problems in the face of meagre and erratic rainfall. Five crops provide the range of practicable subsistence products, but their responses to seasons of good and poor rainfall vary considerably. To simplify the presentation, suppose that any one season can be described as either wet or dry (compare Black and White cells in a lattice, p. 66) and suppose furthermore that the dividing point is so chosen that in a long sequence of years the number of wet and dry seasons is equal. Gould (1963) then collected data on the returns per hectare for the five crops under the two 'strategies' followed by nature:

	Nature's strategies	
	Wet year	*Dry year*
Yams	82	11
Maize	61	49
Cassava	12	38
Millet	43	32
Hill rice	30	71

Yams are the single most rewarding crop, but only if the season is wet; a dry year results in a disastrously low output. On the other hand, both cassava and hill rice do better in dry seasons than in wet. If a farmer plants only two crops, which should the two be and in what combination? Solution of this 'game' shows that the best strategy is to grow maize and hill rice, on respectively 77·4 per cent and 22·6 per cent of the planted area (Abler, Adams and Gould, 1971, pp. 486–7).

The solution of a game-theory problem such as the above requires a formal search procedure of the feasible combinations. The relevant algorithm is in fact a version of linear programming, a technique to which we have already referred (p. 131).

All-in-all, game theory is useful as a normative framework for individual allocation choices but appears to be inherently unsuited for solving the problem with which the previous section ended. Admittedly, the individual firm is no longer treated as taking the external conditions as given, and in *formal* terms a solution can be envisaged for a whole system; the *practical* difficulties make a solution unattainable in situations where the players are envisaged as being firms. Only where the game is played against 'nature' can we reasonably expect to get useful normative statements that apply to a whole system, such as the cropping pattern for a region or the best strategy for avoiding the hazards of flooding.

Central-place theory

Of all the theories advanced in the field of geography, most attention has undoubtedly been directed to that propounded by Christaller (1933 and 1966). His ideas were taken up and elaborated by Lösch (1954) and have been reproduced in many works, for example Berry (1967) and Garner's essay of the same year. Indeed, at least three reviews and bibliographies appeared between 1961 and 1970, such has been the volume of work based on Christaller (Berry and Pred, 1961 and 1965; Szumeluk, 1968; Andrews, 1970).

Christaller's concern was with the size, number and distribution of central places conceived as places which provide services of some sort, such as retail shopping facilities and banks. A large central place is one that has many service outlets and performs a wide range of functions. Although an important central place may be associated with a large town and many activities other than the provision of services, no such equivalence is necessarily implied. Inverness and Fort William are both small towns with a quite 'disproportionate' amount of service functions that are used by people from a very large hinterland. Even more extreme are the market centres in Nigeria and elsewhere that are occupied once every few days as part of a regular periodic market network but are otherwise deserted (Hodder and Ukwu, 1969; see also Berry, 1967). Christaller's problem was to devise a spatial arrangement of central places which would minimize the travel costs of the population in gaining access to the services they require. His objective function, therefore, is to minimize consumer travel costs for the system of cities, subject to certain constraints.

To solve this problem, a number of simplifying assumptions must be made, not all of which will be rehearsed here. In the first place, each central function is treated as having a minimum viable scale of activity, which implies some minimum size of market area. Those functions which have a low threshold are provided at all central places, from the lowest order to the highest order: the highest order central place provides some facilities that are elsewhere not available. Christaller assumed that the central places would be as close together as possible, that is, that market areas would just achieve the threshold limits; also, that all settlements must be in the market area of at least one settlement of next higher order; thirdly, he assumed that the market areas for each order in the hierarchy exhaust the whole space. Given a plain with transport costs uniform in all directions (an isotropic plain) the simplest hierarchical arrangement derived from these assumptions is as shown in Figure 33.

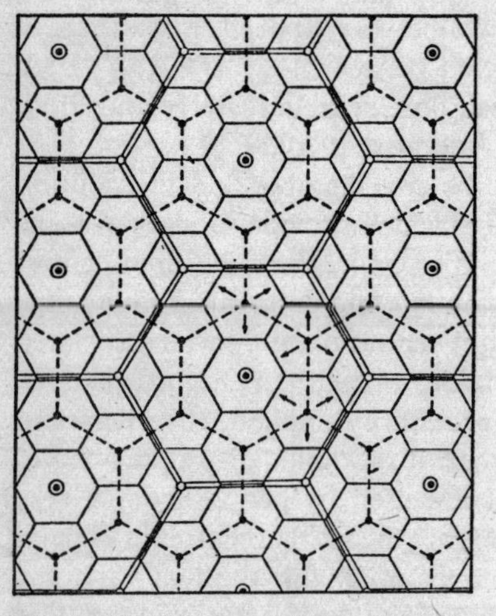

33. Ideal landscape of cities and market areas, after Christaller. (*Source: Garner, 1967, p. 307*)

As Christaller himself pointed out, the arrangement in Figure 33 is only one of several that can be envisaged. The crucial issues are: whether or not one regards a lower-order centre as 'shared' between next higher-order places, as in the K = 3 system illustrated; whether or not the locations of settlements should be adjusted so that the maximum possible number lie on or near routes between the main centres. According to the assumptions made on these two matters, a considerable range of possible ordering patterns can be envisaged, as in Figure 34. Indeed, Lösch adopted the ingenious technique of superimposing the network for one order of central places upon the network for another and then rotating the networks until the maximum degree of spatial association of central places was obtained. The reason for doing this is to minimize the route-distances necessary to link the settlement pattern and hence increase the efficiency of the spatial organization. In this manner, Lösch generated a pattern of sectors radiating from the highest-order central place, city-rich regions alternating with sectors containing few central places.

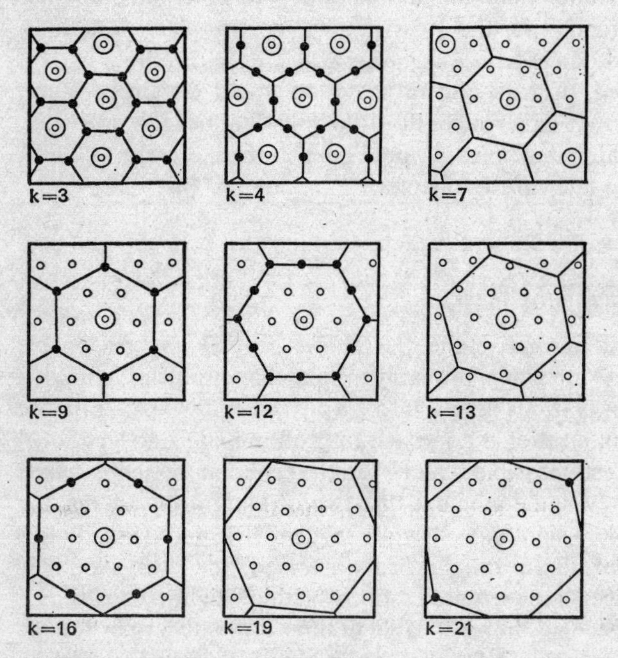

34. Nine smallest hexagonal territories in a Löschian landscape. (*Source: Haggett, 1965, p. 119*)

In the simplest case of K = 3 (Figure 33) the number of central places will run from one highest order, 3 second order, 9 third order, through 27, 81, etc. The progression clearly differs for each K-value, which in turn depends on the variations in assumptions made (see Figure 34).

The framework which has been sketched above provides a logical basis for supposing that a hierarchy of central places exists. Presented in these formal terms, however, the assumptions are very strict and manifestly divorced from reality. At this juncture, however, let us note just two of these assumptions. In the first place, a hierarchy of central places with discrete orders implies that the trade areas for various bundles of services are identical. This may be a useful assumption for theory building but is unlikely to match the realities of the actual world. The second assumption to note is the absence of scale economies in the provision of central-place functions. Christaller and others have assumed that, once the threshold market area is achieved, the unit cost of provision would remain constant. To the extent that either or both of these assumptions is not valid, the discrete ordering of places would be disturbed, a point to which we shall return later in this chapter.

Summary on normative theory

This review of normative theories shows beyond any reasonable doubt that there are grave, probably insuperable, problems in linking theories at the various levels into an articulated whole. For the individual firm, a rational approach to profit-maximizing locational choices is perfectly feasible if all else in the economic system is taken as given. However, the moment that two or more firms are making their own choices simultaneously, or sequentially, and these choices have reciprocal effects on the firms involved, then very complex decision choices are required. Even within highly constrained situations, it becomes impracticable to offer normative rules for the single firm. From this it follows that normative statements for the whole economic system cannot be derived from the micro-economics of the firm. To make any statements applying to the whole space-economy, it is necessary to assume that the public 'good', however this may be defined, will be accepted by the firms. For this to happen, some mechanism other than the competitive market system must be envisaged. In effect, as I have argued elsewhere (Chisholm, 1971a), it is probably necessary to conceive of normative theory for the system of settlements as a guide for the planner, who, through the operation of controls, can impose his view upon the real world. Failing this, the main use of normative theory at this level is to provide a yardstick against which to judge the efficiency of the real world, a task that in itself is hard enough.

POSITIVE THEORIES

The fundamental purpose of positive theories is to account for some empirically observed regularity or set of associations. Thus, the procedure is to take a set of observations and develop a system of relationships that can be regarded as an adequate explanation of the recorded data. In principle, the theorist will usually make his theory as determinate as possible, that is, identify the nature of the causal chains, the direction of causation and the magnitude of the effects (see p. 122). Ideally, the independent variables selected will account for all of the variation in the dependent variable, in which case the system is fully determinate and completely accurate predictions can be made of the magnitude of the dependent variable for any given combination of independent variables. This ideal situation, conforming to the canons of the scientific method, is usually not achieved in the social sciences. Some element of variation in the dependent variable is usually unaccounted for. This variation may be treated as the omission of some relevant explanatory factor or factors, or as a random or chance element in the system.

This last point leads to a more general idea. In any functional system, whether animate or inanimate, there will be an element of chance in all events. A rational approach to developing explanatory theory must therefore seek to isolate the contribution of chance factors and so enable the deterministic part of the system to be identified and specified.

Furthermore, any given phenomenon may occur for two or more quite different reasons. Therefore, we are continually faced with the problem not of developing *a* theory but of choosing *which one* of several on offer is in fact the correct one. A trivial example will serve to make this point. Suppose that a freight movement of 10 tonne-kilometres is recorded; this may result from an infinite array of possible combinations of tonnage and distance moved (two examples are 10 tonnes × 1 km and 1 tonne × 10 kms). There is no *a priori* basis on which to choose the correct combination to explain the observed freight movement and further information is needed. The principle involved in this trivial case has a universal application, as is apparent in the next section.

Urban rank–size relationships

One of the more famous empirical regularities in geography is the urban rank–size relationship, to which we have already referred (p. 87). Following our discussion of central-place theory, it is natural to turn to the ideas of Christaller for a possible explanation. However, the immediate logical problem is that central-place theory postulates discrete classes of urban centres, whereas the rank–size relationship generally assumes the form of a continuous distribution.

Several authors have examined this problem and the following adaptation of central-place theory has been suggested (Beckmann, 1957–8; Mills, 1972). Recollect that central-place theory assumes that the various goods and services supplied from a town all have an identical threshold market area and that there are no economies of scale in their provision. These assumptions are not realistic. Therefore, there is likely to be some variation about the 'correct' size of town at each level in the hierarchy. This variation can be represented as a random variable incorporated into the central-place system, to yield a continuous distribution that matches observed rank–size relationships.

An alternative approach is to propose that if x represents any phenomenon, in this case city size, then x will be normally distributed if it is the resultant of a very large number of other variables that are themselves distributed in a normal manner. (The normal distribution is shown in Figure 35.) Therefore, a normal distribution of city sizes would result from the random operation of many variables. This idea can be extended by proposing that the probability of a given proportionate change in size is the same irrespective of the absolute size of a city. We may then imagine that cities are 'born' at some minimum, or threshold, size and engage in a random walk along a line representing city size and drawn as a logarithmic scale. For each time interval, the chances of each city becoming larger or smaller by a given proportionate amount are the same irrespective of its position on the line. This proposition is equivalent to saying that the chances of moving a given distance along the line are the same at all points. The reason for this equivalence is that on a logarithmic scale a given absolute distance represents the same proportionate change at all points on the scale. By specifying certain rules for the random walk, which involves both increases and decreases in city size, it is quite easy to generate a lognormal distribution, that is, a distribution that is normal when the *x*-axis is drawn in logarithms (Figure 35).

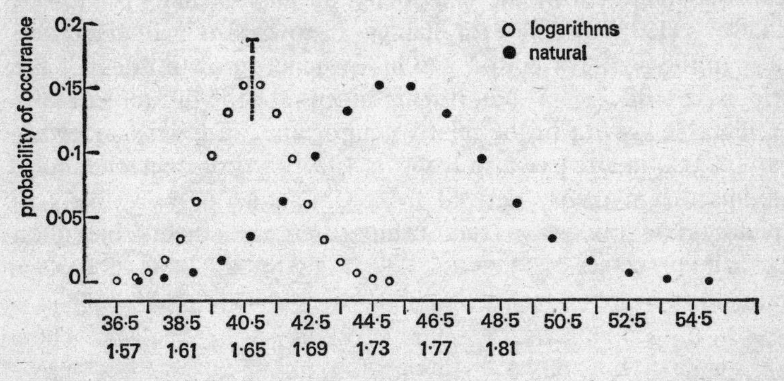

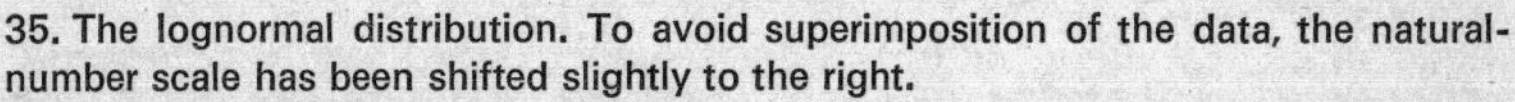

35. The lognormal distribution. To avoid superimposition of the data, the natural-number scale has been shifted slightly to the right.

Now the reader might well ask: what has the lognormal distribution got to do with the rank–size relationships of cities? Referring again to Figure 35, the section of the lognormal curve to the right of T is such that when it is plotted on double logarithmic paper, that is, with both axes in logarithms, it approximates to a straight line. Therefore, T can be regarded as the threshold below which the rank–size relationship does not hold. This idea is entirely consistent with the empirical evidence, since several workers have been quite explicit that rank–size relationships hold only above a certain size of city (for example, Allen, 1954, and Haggett, 1965).

Thus, we have a pretty dilemma in choosing between the Christaller-based and the stochastic theory for urban rank–size relationships. The former probably has greater intuitive appeal, which would be supported by the implausibility of a key assumption in the stochastic model. This is the assumption of equal probabilities of a given proportionate change in size at all points on the size-continuum. This implies that there are no economies or diseconomies associated with city size. Though the empirical evidence is open to challenge (Richardson, 1973a and b), it does seem likely that there are urban scale economies. At least, below about 100,000 population it appears that unit costs of constructing and servicing a town increase with lesser size; and above about 1·0 million, there is evidence that unit costs rise with size. Thus, it seems implausible to suppose that the rate of growth is unaffected by the size of city, an implausibility reinforced by Lever's (1973) evidence for changes in town size in England and Wales in the post-war period. On his evidence, growth rates for the various sizes of cities would, if maintained into the future, lead to a considerable growth in the relative importance of towns somewhat over 100,000 in size (see also Robson, 1973b). Altogether, therefore, it seems that a theory derived from Christaller is more likely to explain urban rank–size relationships than is a theory based on stochastic processes.

Gravity models

Many workers have reported striking distance-decay functions for various features of spatial interaction. Further examination of these regularities, especially when consideration is given to many origins and destinations simultaneously, suggests a close analogy with Newton's law of gravitational force, F_{ij}, between pairs of heavenly bodies:

$$F_{ij} = a\frac{M_i . M_j}{d_{ij}^2}$$

where M is the mass of a heavenly body, i and j represent two such bodies, d_{ij} is the distance which separates them, and a is a constant. In other words, the gravitational attraction is positively related to the product of the two masses and inversely to the square of intervening distance. In the gravity model analogy, the masses M_i and M_j can be represented by some measure of the economic size of cities, such as total population or purchasing power; and, as we have seen in Chapter 2, distance (d_{ij}) can be measured in route kilometres, trip time, trip cost, etc. Thus, the above equation can be rewritten as:

$$T_{ij} = k\frac{O_i . D_j}{d_{ij}^2}$$

to give an estimate of the volume of interaction (T_{ij}) between any pair of origins and destinations.

This simple gravity model suffers from a number of defects. In the first place, there is no obvious reason why the power of distance should be 2, and instead of writing d_{ij}^2 it is usual to use the more general term d_{ij}^{β}, with β to be derived empirically. In the second place, whereas a planet or a star can simultaneously exert a gravitational attraction of the same magnitude on any number of bodies without 'using up' the supply of gravity, human interaction between one place (O_i) and another (D_j) must be thought of as an alternative to interaction with other potential destinations. Therefore, each origin is treated as having a defined supply of originating trips, and each destination a similarly given number of trip ends. Consequently, the fully specified gravity model includes additional terms, sometimes called balancing factors, which ensure that all originating trips are cleared and all terminating trips supplied, with neither excesses nor deficits anywhere in the system (Wilson, 1970). It then becomes a matter of iterative calculation to obtain the gravity-model parameters that give the best fit to the empirically observed data. However, in practice the parameters so derived vary from case to case; also the appropriate metric both of distance and mass, but especially the latter, also varies. Notwithstanding these reservations, gravity models have performed remarkably well in many studies.

Gravity models have been applied in many contexts, from studies of the patterns of international trade through inter-regional trade flows to the problems of urban planning and retail location. Yeates (1969; see also Johnston, 1973) examined the trade flows of six countries in 1964 and found that the correspondence between predicted and actual trade volumes, measured by the value of r^2, varied from 0·26 for the United Kingdom to 0·51 for Canada. The

United Kingdom's foreign trade was at that time much affected by Commonwealth preferences and when a term was introduced into the gravity-model equation to take account of this fact, the model then yielded an r^2 value identical to that for Canada. On this revised basis, France then emerged as the country for which the model performed least well, only 24 per cent of the variance being accounted for. Yeates' study represents an advance over previous work by economists (Beckerman, 1956; Linneman, 1966), in that he incorporates into his model the effects of trade blocks, which to some extent disrupt the pattern of transactions that would occur 'naturally'.

Substantially better results than the above were obtained by Chisholm and O'Sullivan (1973) for the intranational distribution of freight flows in Great Britain. Using road freight traffic only, recorded by seventy-eight origin–destination zones, and experimenting with various values for the parameters, the lowest value of r^2 for estimated flows in relation to actual was 0·73 and the highest 0·80. Rail freight proved to be much less amenable to a gravity-model formulation, and disaggregation to the eleven main commodity groups was also notably less successful than when total freight flows were used.

Better still were Keeble's (1971) results for the spatial distribution of migrant firms in Britain, though the small number of geographical areas available to him required the use of rank–order correlation techniques, which are less powerful than regression methods. For firms mobile in the period 1945–65, he fitted gravity models to the data for the three main source areas, using a variety of distance exponents. Of the best fits so obtained, the lowest yielded a Spearman's rank correlation (r_s) of 0·814 and the highest 1·00.

Other workers have obtained very encouraging results in applying gravity models to flows of migrants between regions (Hart, 1970; Masser, 1970). Masser calibrated gravity models with migration data obtained from the 1961 and 1966 censuses for the six conurbations of England and Wales and found not only that very good fits could be obtained but also that models calibrated with 1961 data gave excellent predictions of the flows recorded five years later. Furthermore, a simple gravity-model formulation appears to fit the data just as well as more sophisticated formulations that incorporate economic variables such as wage rates, differences in which one would expect to affect migration patterns.

Migration patterns in the United States have been examined by Schwind (1971), using information from the 1960 census, which recorded migration that had occurred in the previous five years. The data were aggregated in two ways: 225 regions were identified as functional units on the basis of commuting patterns; a rather coarser mesh of reporting areas, numbering 133 in total, was also used. Schwind used an unconstrained gravity model and found that in the simplest version 55 per cent or 57 per cent of the gross migration flows could be accounted for, according as the smaller (225) or larger (133) geographical areas were used. In this simple model, the only variables employed were the populations of the areas of origin and destination and the intervening distance. By introducing further variables describing the zones of origin and destination, the performance of the gravity model was raised to 72 per cent and 77 per cent respectively. Eleven such additional variables were used for both origins and destinations, being variables related to the propensity for out- and in-migration, for example, degree of urbanization and level of unemployment.

Gravity models have also been extensively used in retail trade studies since about the middle of the 1960s (Cordey-Hayes, 1968; Lanchester Polytechnic, 1969; Davies, 1973). In this particular context, two versions of the gravity model have been developed. The first derives from Reilly's (1929) law of retail gravitation, which proposes that the frequency with which residents in an area shop in two competing towns is proportional to the towns' respective sizes and inversely proportional to the square of the distances from the area in question to the two towns. In this way, the boundary between two towns' catchment areas can be identified as the line of equal patronage. An alternative version is designed to establish the trade boundaries of centres on the premise that either side of a line of indifference (the boundary) customers will exclusively use one centre in preference to the other. The first version is probably the more realistic.

Taking the first version as the point of departure, a further refinement must be introduced. The boundary representing the limits of an urban trade area should be defined as the line where the probability of visiting the trade centre in question is equal to the combined probability of visiting all other centres. This amended version can be stated formally as:

$$P_{ij} = \frac{S_j/D_{ij}^{\beta}}{\sum_{=1}^{n} S_j/D_{ij}^{\beta}}$$

where P_{ij} = the probability of an individual located at i interacting with the central place j
S_j = the size of the central place j
D_{ij} = the distance from i to j
β = an exponent

Huff (1973) used this technique to obtain the spheres of influence for seventy-two cities in the United States (Figure 36 shows his results).

The intuitive rationale for the gravity model is obvious in Zipf's (1949) use of the term 'least effort', that is, mankind so arranges his affairs that within the relevant constraints he minimizes the time/cost/trouble of overcoming distance. Put more formally, man may be regarded as a rational being who at least attempts to maximize his utility. In this wise, therefore, a distance-decay function may be treated as a spatial demand curve; the greater the distance (higher cost), the less the interaction (or demand) (see p. 113).

However, this suggestion conflicts with some empirical evidence on the relationship of distance to transport costs. For road freight in Great Britain, the level of terminal charges to haulage charges is such that the former are equivalent to a haul of 127 kilometres. As the average distance freight moves is under 48 kilometres, terminal charges amount to between two-thirds and three-quarters of road haulage charges. Therefore, despite the fact that transport costs amount to at least 9 per cent of total production and distribution costs for manufacturing industries, the spatial variation in transport costs is very small (Chisholm, 1971c). Despite this, we have already seen that a gravity model can be fitted to the inter-zonal flow data with considerable success. Consequently, if the gravity model β coefficients are to be interpreted as spatial-demand curves,

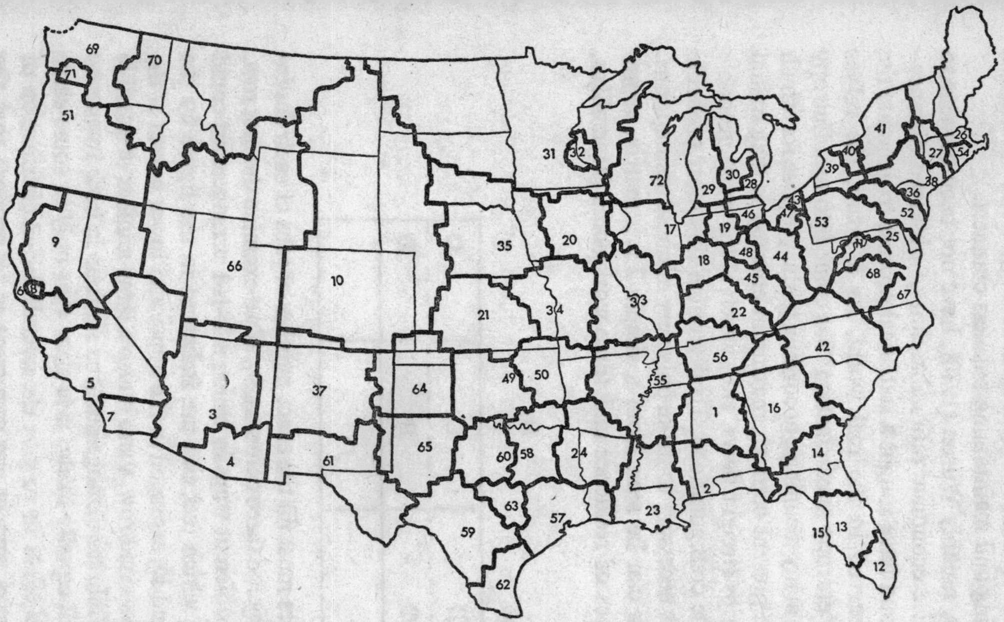

1. Birmingham
2. Mobile
3. Phoenix
4. Tucson
5. Los Angeles
6. San Francisco
7. San Diego
8. Oakland
9. Sacramento
10. Denver
11. Washington, D.C.
12. Miami
13. Tampa
14. Jacksonville
15. St Petersburg
16. Atlanta
17. Chicago
18. Indianapolis
19. Fort Wayne
20. Des Moines
21. Wichita
22. Louisville
23. New Orleans
24. Shreveport
25. Baltimore
26. Boston
27. Springfield
28. Detroit
29. Grand Rapids
30. Flint
31. Minneapolis
32. St Paul
33. St Louis
34. Kansas City
35. Omaha
36. Newark
37. Albuquerque
38. New York City
39. Buffalo
40. Rochester
41. Syracuse
42. Charlotte
43. Cleveland
44. Columbus
45. Cincinnati
46. Toledo
47. Akron
48. Dayton
49. Oklahoma City
50. Tulsa
51. Portland
52. Philadelphia
53. Pittsburgh
54. Providence
55. Memphis
56. Nashville
57. Houston
58. Dallas
59. San Antonio
60. Fort Worth
61. El Paso
62. Corpus Christi
63. Austin
64. Amarillo
65. Lubbock
66. Salt Lake City
67. Norfolk
68. Richmond
69. Seattle
70. Spokane
71. Tocoma
72. Milwaukee

36. United States: seventy-two urban spheres of influence. (*Source: Huff, 1973, facing p. 305*)

demand is related less to costs than to some other consideration related to distance – perhaps the perception of opportunities and the ease of establishing and maintaining business contacts.

Other workers, notably Wilson (1970), have not been content to accept the intuitive economic rationalization of observed distance–decay functions and have sought a statistical process to generate the distributions observed in the real world. As a result, the rather forbidding term 'entropy maximization' has come into use. Entropy is a concept with many meanings, according to the purpose for which it is used. In the present context, maximum entropy is that spatial distribution of trips (freight flows, etc.) which is the most probable of all the possible configurations. To determine this most probable configuration it is necessary to make only very minimal assumptions.

The basic idea can be seen from a simple 2 × 3 matrix, representing two places of residence and three workplaces (Macmillan, 1973):

			D_j	
		2	3	2
O_i	4	AB AC	C B	D D
	3			

The seven workers must fill the exact number of jobs at each destination, indicated by the three columns D_j. If we examine the first row, there are twelve different ways in which a 2–1–1 arrangement could be generated, of which just two are indicated in the figure by the letters, A, B, C and D, arranged in the upper and lower halves of the first row. For the second row, there is also a large number of possible arrangements. While the arrangements of workers in both rows may be regarded as all equally likely, it is clear that only in some cases will the *combined* effect be to give the overall 2–3–2 distribution of trip ends. The next step in the argument is to propose that the probability of the occurrence of any distribution of trips in the whole matrix conforming to the aggregate flows is dependent upon the number of states that is associated therewith, that is, the number of combinations of A, B, C, D, etc. in the matrix cells. The larger the number of such states, the more probable is that distribution.

Clearly, the whole operation becomes immensely complex to handle, even with a small matrix as used above. However, the underlying idea is quite simple and from it can be derived a distribution which is the one most likely to occur. And the interesting point is that a distance-decay pattern is the end result, a pattern that conforms to the empirically observed regularities.

Satisficer models

The concept of man as a rational optimizer who, possessed of full knowledge, seeks and achieves the absolute best in all the choices he makes has been criticized in many quarters. Two rival concepts have been advanced as better representations of the way in which actual choices are made, whether these be in a spatial or non-spatial context; both are illustrated in Figure 37. Imagine that a firm must choose a location and that objectively it can be shown that potential profits vary between locations as shown by the solid line *a* in the figure. An omniscient manager who seeks to maximize his profits would choose the point of inflection, that is, location *X*. If we follow Simon (1957), then we would suppose the decision-maker to have

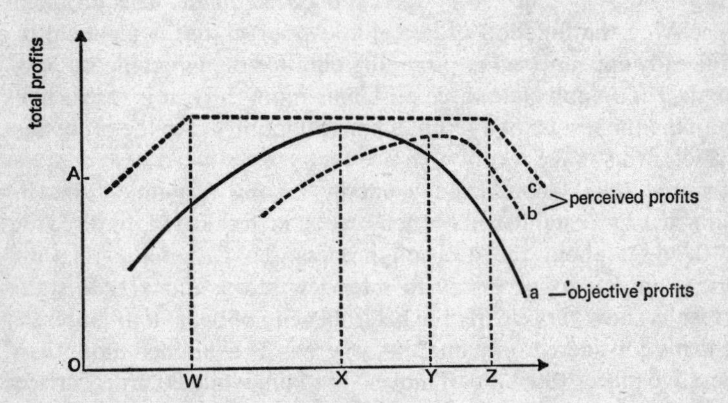

37. Optimizing, bounded rationalist and satisficer behaviour.

incomplete information but that within the knowledge available he tries to be rational. With this notion of 'bounded rationality', the perceived situation might be as depicted by curve *b*. As this has a point of inflection, the rational *choice* will be at location *Y* even though this is not objectively the optimum location. Finally, curve *c* represents another form of perceived opportunity, related to the idea of 'possibilism' already discussed (see p. 40) and also Rawstron's (1958) idea that instead of seeking the optimum location attention should focus on the boundaries beyond which locations are not feasible. In this third version, locations from *W* to *Z* inclusive are perceived as equally profitable; the decision as to which one will in fact be chosen then depends on other factors, such as the pleasantness or otherwise of the local scenery. The locations from *W* to *Z* are all regarded as satisfactory.

Hence, a 'satisficer' may be conceived as someone whose perceived opportunities are as *c* in Figure 37 and who is essentially indifferent to locations between *W* and *Z*. Alternatively, a satisficer is someone who defines a satisfactory level of profit, say *OA* in the figure, and then, irrespective of the shape of the curve of opportunities, seeks a location that meets this minimum requirement.

The fundamental problem with this approach is how to specify the nature of the perceived opportunities. These are necessarily different from the objectively available opportunities and are not necessarily closely reflected by the overt choice made. This problem of specifying the function of perceived opportunities is identical to that already encountered in providing data for the game-theory pay-off matrix. To appreciate these problems more fully, it is instructive to examine the results of a recent empirical inquiry into the processes by which firms make location choices.

Townroe (1971) conducted a survey among fifty-nine firms in Britain that had established branch plants, or re-located, to discover how they set about the decision process. To gather the relevant information, it was necessary to interview senior managerial staff. The results show very clearly the heterogeneity of behaviour patterns. Five firms considered only one site, whereas 24 examined more than 10 possible places to which to move. Or again, while 20 firms carried out an economic evaluation of several sites, 13 made no assessment at all and the remaining 26 evaluated only one – the site actually chosen. Despite the thoroughness of the inquiry, the behaviour of the firms can be summarized only in a rather qualitative way, as:

Small firm		*Large firm*
Chance knowledge of site	⟷	Systematic search
Local search	⟷	National search
Personal contacts	⟷	Official agencies
Few sites considered	⟷	Many sites considered
Satisficer	⟷	Optimizer

Furthermore, can the results from fifty-nine firms be treated as representative? Probably not. Thus, an immense investment of effort is required to provide useful evidence on the nature of the perceived world and of the satisficer–optimizer dimension of real behaviour.

Perception studies

There are several strands of thought which converge on the problem of how man actually perceives the world around him. In Chapter 2, the argument was developed that regions do not exist as objective realities independent of the observer but are mental constructs that vary both with the person concerned and the purpose for which they want to identify regions. This point is reinforced in Chapter 3, by the interrelationship of fact and theory via philosophy: what we recognize as facts is conditioned by our prior theoretical constructs. Then, as we have just seen in the preceding section, when we move from the optimizing model of human behaviour to the satisficing standpoint, problems arise as to how individuals and firms set about the task of searching the environment to construct their schedule of perceived opportunities. Thus, we must face up to the issues raised by the explicit incorporation of perception processes into positive theorizing.

Figure 38 presents a very simple formulation of the role of perception. The external, objective environment consists of all the elements of the natural and the man-made world relevant to a particular situation. This objective world is to be regarded as a reality that is independent of any person or group of people. It is this real world about which we get messages directly through our senses – of seeing, hearing, smelling and touching – as well as indirectly from other people, by the spoken word or in writing. The world so perceived may bear a close or a distant relationship to the real, objective world.

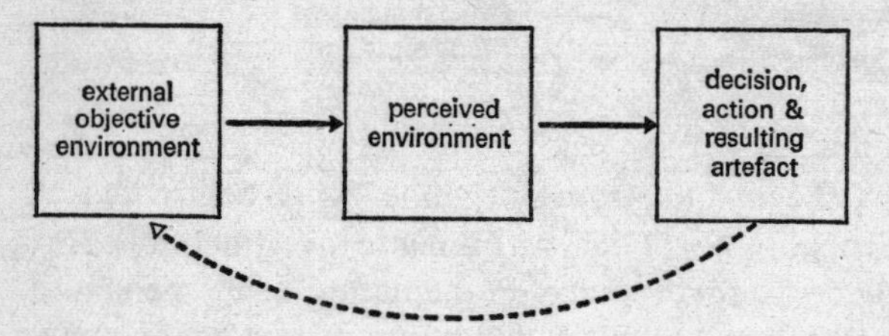

38. The place of perception studies.

Now it is abundantly clear that we cannot fully understand human actions unless we know what are the mental constructs that men have about the environment. Nor can these constructs be inferred with confidence either from our own assessment of the objective environmental conditions or from the actions of mankind. Some kind of direct probing of the middle box in Figure 38 is essential, and is a task that has been attracting a good deal of research interest in recent years (Brookfield, 1969; Downs, 1970).

As an area of research inquiry, perception studies are both exceptionally difficult and still in a relatively primitive stage. Indeed, one could say that the task of devising suitable measures of perception has hardly begun; certainly we are not anywhere near formulating theories or laws about perception. Nevertheless, the venture is necessary to a proper understanding of actual behaviour; the study of perception is an essential ingredient of positive theory. A well-known example of perception surfaces is the study by Gould and White (1968) of the attitudes held by school-leavers towards various parts of Britain as places to live. Selecting schools in various cities, the children were asked to assign to specified locations scores, on a scale desirable–undesirable, from which generalized values could be obtained and mapped. Figure 39 presents the resultant image from the vantage-point of Newcastle-upon-Tyne; high scores represent

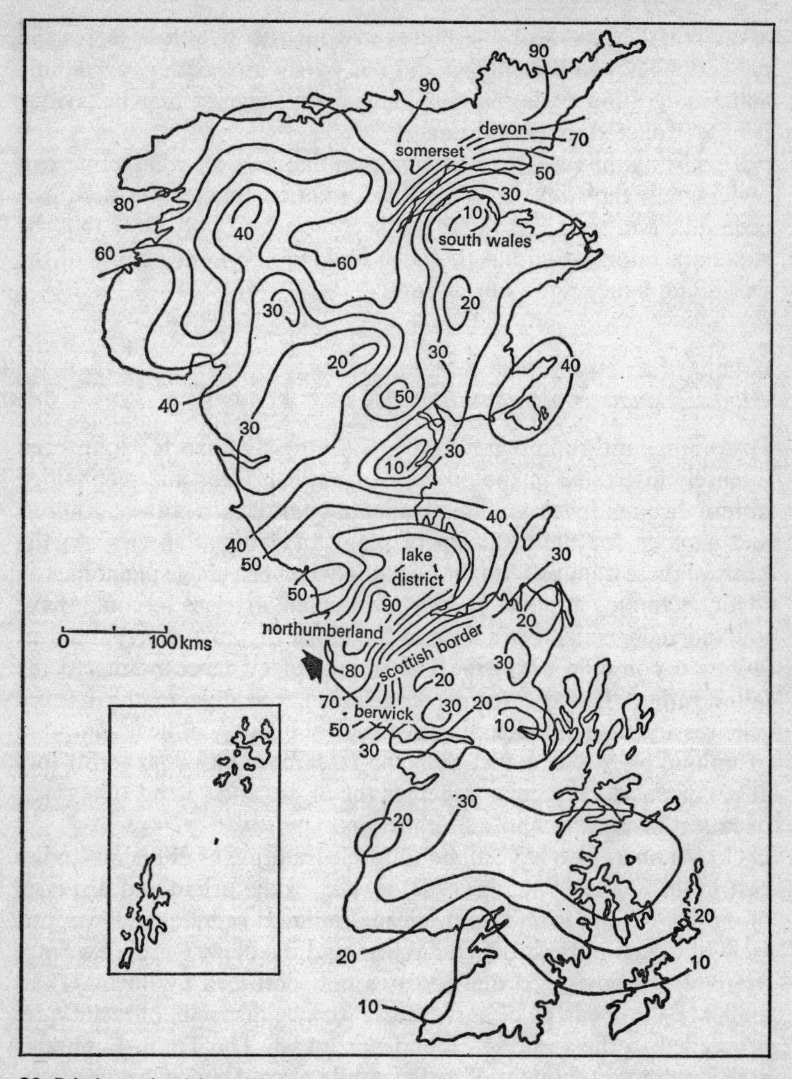

39. Britain: school-leavers' perception surface, Newcastle-upon-Tyne. (*Source: Gould and White, 1968, p. 169*)

areas regarded as desirable places in which to live, low scores the converse. Though the authors did not pursue the matter to examine actual migration patterns, it is plausible to suggest that behaviour is in fact related to mental images.

This being the case, the next question that arises is whether we can understand the way in which information passes through the economic and social fabric. It has become abundantly clear that the supply of information has a crucial bearing upon our images of the world and hence upon our actions.

Diffusion studies and information flows

Historians, anthropologists and archaeologists have for long been intensely interested in the processes by which ideas and technology diffuse through space, an interest that has parallels in botany, zoology and geology for the diffusion of plant and animal species. At the heart of these inquiries lies the question whether a single phenomenon – for example, a plant species or human artefact – could have evolved independently in several locations or must be regarded as having a common ancestry. The weight of evidence points to the latter rather than the former conclusion, certainly in the natural sciences and also in much of man's ancient history. If the assumption of unique origins is made, then the researcher has a powerful tool at his disposal for piecing together the origin and spread of species, technical inventions and cultural traits.

Geographers also have an honourable tradition of diffusion studies cast in this mould. Sauer's (1952) classic on the origin and dispersal of cultivated plants and domesticated animals represents a synoptic view of events in both the Old World and the New. On the evidence available, he envisaged that the area now occupied by Burma/Thailand was the 'hearth' of agriculture, whence dispersal of techniques proceeded to the west, south and north-east. The Tigris–Euphrates area, commonly thought of as the cradle of western civilization, was but a secondary, or derivative, centre – nonetheless important for western civilization but not quite the original source of agricultural innovation that it is sometimes portrayed to be (Figure 40).

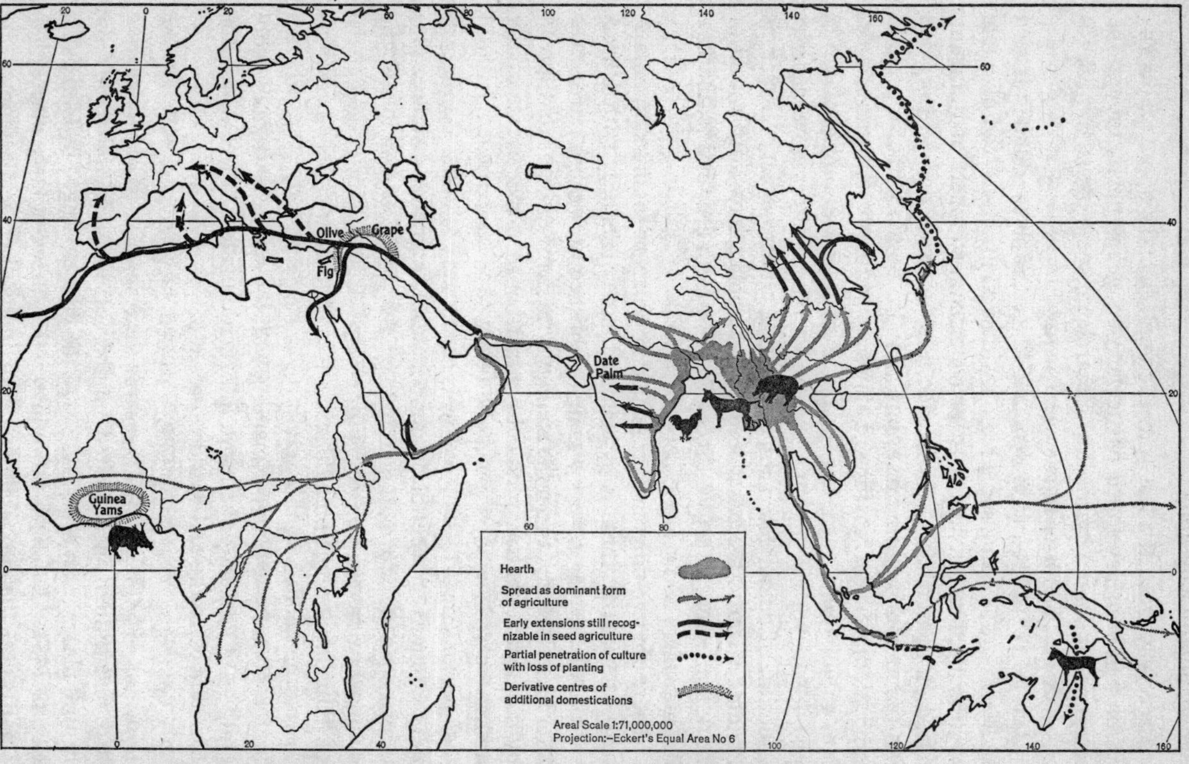

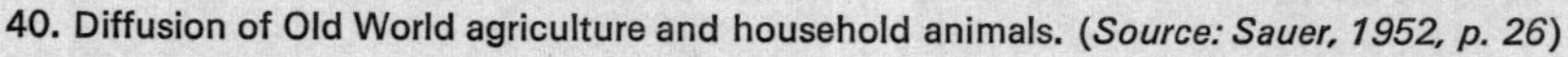

40. Diffusion of Old World agriculture and household animals. (*Source: Sauer, 1952, p. 26*)

Especially since the last war, a complementary approach to diffusion has been developed particularly by geographers, though the enterprise is shared by medical researchers, sociologists and economists. A much shorter time scale is involved, perhaps a few decades instead of centuries and millennia. The geographical domain is also restricted, often being some region within a nation, instead of whole continents. The attempt is to provide *formal* models of diffusion, based on some theory of the processes involved, that will provide both a good description of events and also be the basis for making valid predictions at least in the short or medium run. For example, how rapidly and in what geographical manner will family planning practice in Nigeria and India spread? What factors governed the adoption of colour television sets within advanced nations such as Britain? Clearly, the possible practical applications of this kind of work are legion, a fact well recognized by farm economists, marketing managers and many others.

There are two different but related approaches to diffusion modelling. The first of these is associated most particularly with Hägerstrand, whose major work was first published in Swedish only a year after Sauer's study. From the analogy with many diseases that spread by contagion, that is, physical contact between the disease-carrier and the person who is given the infection, Hägerstrand envisaged that persons near to someone who has already adopted the innovation are more 'susceptible' than persons at a greater distance. This is plausible if information is passed by word of mouth and the demonstration effect of seeing for oneself. Consequently, the pattern of diffusion depends on a distance-decay function closely analogous to that embedded in the gravity model (see p. 150). The problem then becomes one of experimenting with the spatial pattern of probabilities of contact – the mean information field – to see which one best replicates the real-world processes.

A complementary view that was also suggested by Hägerstrand (Hudson, 1969; Berry, 1972) regards the transmission of information as concentrated on a limited number of channels which link the central places of an urban system of the kind envisaged by Christaller (see p. 143). As there is more interaction between large places than small ones, the probabilities of an innovation being adopted are higher in big cities than in little ones. Thus, one may envisage that an innovation filters down through the hierarchy of urban places: while London and Paris are *au fait* with the latest fashions, Bristol and Bordeaux will be laggard in adopting the latest style. Certainly the date of first opening of television stations in the United States between 1940 and 1968 conforms to this pattern (Figure 41).

However, Hudson asks the question: can either the neighbourhood or the hierarchical process generate the familiar logistic curve which has been widely observed to describe the accumulated number of innovation adopters? This logistic, or s-shaped, curve reflects the slow initial response to an innovation, an accelerating tempo of adoption and then a slackening in the pace as saturation point is approached. Hudson shows that in fact such an s-shaped curve requires both the neighbourhood, or contagious, process and hierarchical filtering to be operative simultaneously, though their proportionate importance varies with the stage in the diffusion process.

Interest among geographers in this and kindred problems of diffusion processes has been great and the resulting literature is now substantial (L. A. Brown, 1968a and b; Brown and Moore, 1969; Morrill, 1970; Gould, 1969b). The main thrust of work is to identify which kinds of diffusion process operate at what geographical scales and over what time periods, to see just how general the observed regularities are and for what classes of phenomena (Brown and Cox, 1971). One of the major research areas that is opening up as a consequence concerns the ways in which information actually gets passed from individual to individual, from firm to firm.

Much of the pioneering work on contact systems has been done in Sweden, especially by Törnqvist (1970; Pred and Törnqvist, 1973), who has developed the interesting idea of contact potential. Through careful study of the work patterns of various kinds of workers, especially in tertiary employments – senior management, data control, clerical, etc. – he has identified those occupations which require a great deal of personal contact with colleagues outside the plant or office in which they work. He was able to calculate the amount of such contact potentially available at the main centres of population. This he did by estimating the number of contact-intensive jobs at each place and the amount of working time there available in a day

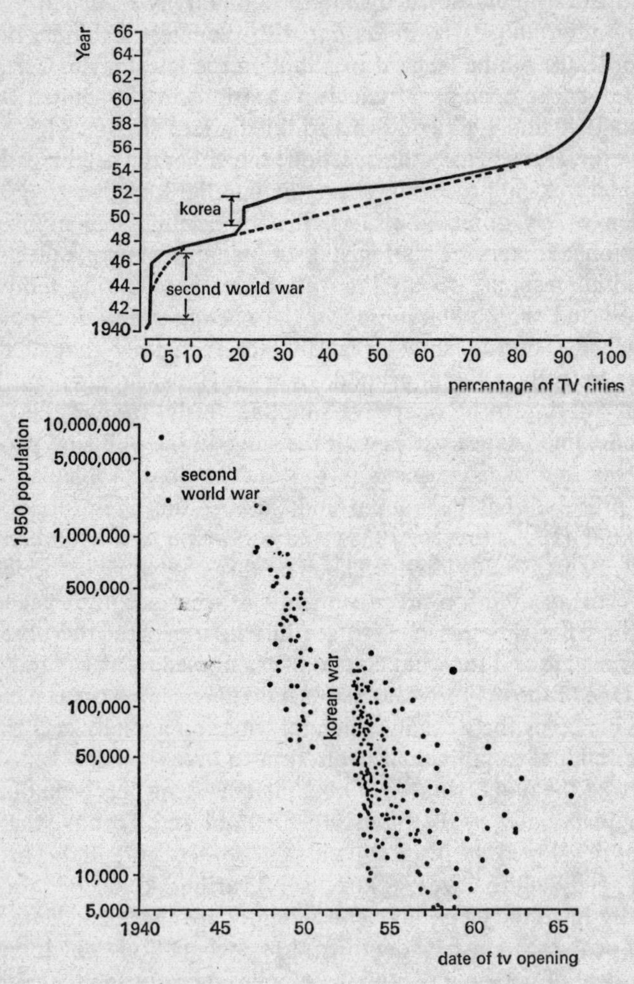

41. United States: diffusion of television. (*Source: Berry, 1972, p. 119*)

if the return journey were to be made within twenty-four hours. Contact potential is analogous to the idea of population potential (see p. 84), though only a specially defined part of the workforce is taken into account and a particular measure of 'distance' is used which takes account of both frequency and speed of travel by the various modes, singly and/or in combination. His map of contact potentials within Sweden is extremely revealing and offers much insight into the process of urban growth in that country. As Figure 42 shows, Stockholm has the greatest contact potential, but though there is a general tendency for values to decline with distance from the capital, Göteborg and Malmö on the south-west coast prove to be surprisingly well placed. Clearly, the spatial distribution of the tertiary sector, and especially of administrative jobs, is likely to be affected by the contact potential values, which must count as an important factor affecting regional rates of economic growth.

CONCLUSION

However one may define a subject, intellectual respectability seems to require that a discipline has its own indigenous corpus of theory. In this respect, geography has been somewhat lacking, and as recently as 1967 and 1969 Harvey felt it necessary to argue that one of the major tasks ahead is to develop geographical theory.

Theories of the location of economic activity derive from the work of non-geographers, mainly economists – von Thünen, Launhardt, Weber, Lösch, Isard, Greenhut and others. Christaller (1933 and 1966) stands out as the exception who is a geographer, and central-place studies are arguably the geographer's greatest contribution to theory. Certainly there has been a quite extraordinary volume of work published by geographers exploring the concept of urban hinterlands and the hierarchical organization of urban functions (Dickinson, 1947; Gottmann, 1961; Berry, 1967). Perhaps the other area in which geographers have made an outstanding contribution is in the study of diffusion, following in the steps of Hägerstrand in particular. Other theoretical constructs are generally derivative in nature, such as the gravity model of interaction derived from Newtonian physics;

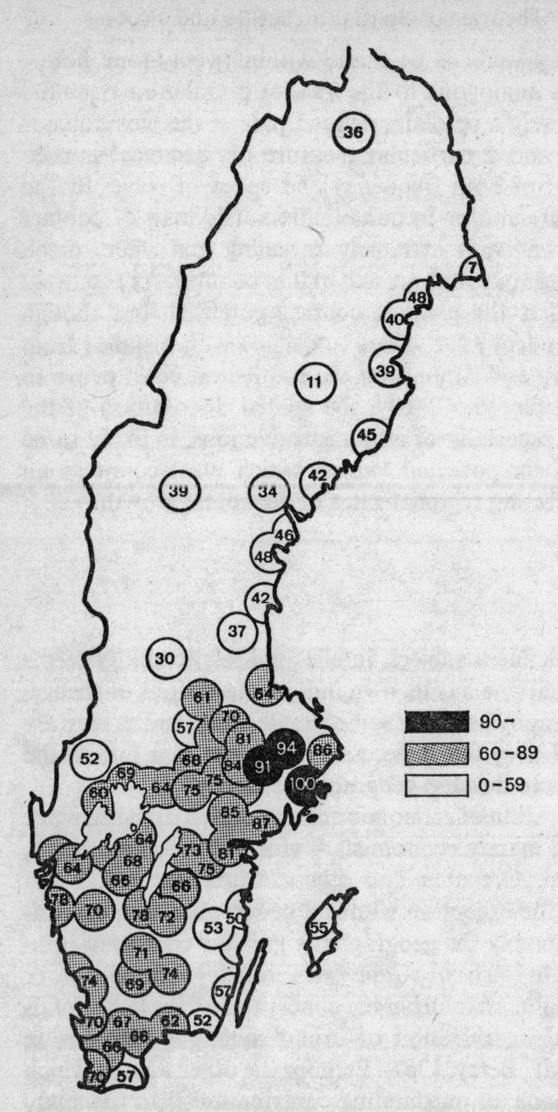

42. Sweden: 1970 contact potentials (Stockholm = 100). (*Source: Pred and Törnqvist, 1973, facing p. 96*)

migration models developed from Ravenstein's (1885) ideas on the minimization of distance moved; concepts of regional specialization and consequential trade relationships taken directly from economics; and so on. However, the 1960s marked a distinct shift in emphasis such that geographers have consciously tried to relate their own work to some corpus of theory or have attempted actually to develop theory.

In a chapter such as this, it is impossible to cover all of the theories currently in use by geographers. It will be sufficient if I have shown the considerable richness of theoretical constructs in the subject, both in the normative and the positive fields, and have succeeded in conveying something of the excitement which attends both the construction of theory and, in the case of positive theory, its testing. I also hope that the links between theory and the descriptive and geometrical traditions of the subject will have become abundantly clear. After the hiatus in the period approximately 1920–50, when many geographers retreated to a descriptive treatment of unique events, the subject has emerged to a position of much enhanced strength in which we may look forward to a healthy mixture of empiricism and theory. But what else is there in store for the subject? That is a question to which the next chapter is devoted.

6. Whither Next?

Has geography experienced a 'revolution' in the last decade or two? Is the subject now identifiably different from what it was in the recent past? These are two of the fundamental questions posed at the very beginning of this book. It is now time to attempt an answer and to look forward to the unfolding future to see whether we can discern 'the main chance of things as yet not come to life'. Two dominant themes have become apparent in the earlier chapters, the opposed themes of continuity and of change. What has been the balance between them?

Continuity is manifest in the geographer's abiding interest in the geometry of space, initially terrestrial geometry but more recently the geometry of other kinds of space. Part and parcel of this concern is the general problem of transforming space from one form to another, questions of locating phenomena in space, the spatial classification of phenomena and the identification of relationships between objects and events in a spatial framework. This geometrical tradition is manifest throughout the last five centuries and its vitality remains to the present day.

Allied to this fundamental trait of geographical research is an equally long tradition of descriptive studies. Description used to be regarded as an end in itself, especially in the context of regional geography: the purpose was to convey an accurate impression of what a place is like and what goes on there. In recent decades, description as an end in itself has been at least partially superseded by description as a means to an end – explanation of the observed phenomena. In any scientific enterprise, an element of precise description is essential, if only to enable other workers to replicate an experiment and thereby check the findings. Unfortunately, in geographical circles the fascination of the 'scientific' approach has frequently caused explanation to be elevated to a pedestal far above the facts for which an explanation is offered. *Mere* description is dismissed, with the result that much modern work lacks the kind of insight vouchsafed by earlier work. My own view is that the pendulum has swung much too far in the one direction and that there is a serious danger of geographers forsaking an essential part of their task. The point at issue is the following. Traditional descriptive geography is predicated on one view of the causal system (the environmentalists' view; see p. 37) and therefore conformed to a standard pattern. The need now is for experiment with other frameworks within which to undertake geographical description, in which both the sequence of material and the coverage/emphasis will vary from case to case. Presentation of factual material is thus conditioned by, and contributes to, the explanatory account that is being developed. Two recent regional geographies exemplify this kind of development. Paterson's (1960) study of North America is outwardly arranged on fairly traditional lines but the reader quickly appreciates that the author has selected a limited number of themes and arranged his material accordingly, so that the emphasis varies strikingly according to the part of the continent being discussed. By contrast, Brookfield's (1971) examination of Melanesia adopts a framework that is explicitly related to concepts of central-place and location theories, ideas on 'colonialism' and trade, the change from a subsistence to a cash economy, etc.

The fact that I feel it necessary to urge the virtues of description, albeit of a kind different from the traditional one, marks the fact that a major change in geography has indeed occurred. Although perhaps only a change in emphasis, the magnitude of the alteration is sufficient in itself to constitute a revolution, a revolution that could be described as placing description in the role of handmaiden to explanation. However, before we pursue this theme further, a related development must be noted.

In a very real sense, the 1960s saw a *return* to the canons of scientific inquiry after the emphasis placed upon the uniqueness of phenomena, especially notable in the inter-war period after environmental determinism had fallen into disrepute. The essential point to grasp is that Huntington in particular regarded the environmentalists' stance as a hypothesis to be tested but, in common with other researchers of his day, lacked the tools to pursue this enterprise very thoroughly. Modern statistical methods and related conceptual advances provided the means by which geographers could more rigorously approach the problems that have been their traditional concern. The question above all other questions in the geographical domain remains the relationship of man to his physical environment, a question that is as far from resolution as the age-old conflict of view regarding the respective roles of genetic inheritance and of nurture in forming our individual characters. However, the man/environment issue has not been at the forefront of geographical work in recent decades. The traditional problems to which geographers have put their new skills have been of a lesser order than this, though nevertheless important in their own right. Questions such as the reasons for the pattern of settlements and the localization of employment, changes in space-relationships over time and, as the network of communications evolves, the geographical diffusion of innovations and similar matters have attracted the bulk of recent work in which the new techniques have been applied. Yet such studies may fairly be described as 'traditional' in the sense that the same issues attracted the attention of previous generations of geographers; old problems have been framed in new terms.

Thus we may understand why geographers became so excited during the 1960s as the 'quantitative revolution' progressed, a process correctly identified by Harvey as a more fundamental shift to explanatory modes of thought. The realization dawned that where a qualitative and descriptive approach had previously sufficed, it was now possible to press analysis much further than had previously seemed conceivable. Although the roots of this change can be traced back to the early decades of this century and even earlier, the adoption process undoubtedly accelerated during the 1950s and 1960s. If we take an idea from innovation diffusion studies, we may imagine a logistic (s-shaped) curve of adoption by geographers of the quantitative and scientific approaches. The steepest part of this curve almost certainly lies in the two decades just past; the adoption curve is now flattening out as 'saturation' is reached.

Therefore, in respect of analytical techniques geography has indeed experienced a revolution. However, in terms of the subject-matter studied the position has changed less rapidly – indeed, one might almost postulate a lagged relationship between the two. The 'capes and bays' kind of geography, related to the early days of discovery and exploration, survived in the school curriculum into living memory. Its descriptive successes were traditional regional geography and its counterpart in economic geography, known as commercial geography; in both cases, the emphasis was on description, either of many phenomena in one place or a single (economic) phenomenon distributed over the globe. Both have given ground to a problem orientation, an analytical attempt to explain patterns of phenomena, but, as already indicated, the problems selected have been very largely derived from the subject's past history rather than from contemporary issues. Even here, though, change is evident, as we shall see in the next few pages; perhaps the 1970s will be regarded by future historians as the decade in which the problems geographers discuss caught up with the analytical skills the profession can now boast. It is around this issue, however, that some serious misconceptions seem to exist and before proceeding it is necessary to clear from the path some of the obstructing vegetation.

Misconceived counter-revolution

In the excitement of events, many geographers apparently imagined that the 'quantitative revolution' would open a new millennium of scientific advance. As the realization has dawned that despite great progress daunting problems remain and that the new millennium is somewhat tarnished, some have voiced grave disappointment. The higher the hopes, the greater the dismay. Thus, some eminent geographers have felt moved to urge upon their colleagues the need for a new paradigm, that is, for a new and 'better' approach that should yield firmer answers to questions. In the Preface to *Directions in Geography* (1973), Chorley rightly refers to the contributors to that volume as major figures associated with raising the standards of geographical work but then goes on to remark that the events of the previous ten or fifteen years to which they contributed are now 'ancient history'. This is pitching expectations at a very high level indeed, in the light of which the following passage will occasion no surprise.

> The quantitative revolution has run its course, and diminishing marginal returns are apparently setting in . . . There is a clear disparity between the sophisticated theoretical and methodological framework which we are using and our ability to say anything really meaningful about events as they unfold around us. There are too many anomalies between what we purport to manipulate and what actually happens. There is an ecological problem, an urban problem, an international trade problem, and yet we seem incapable of saying anything in depth or profundity about any of them.
>
> (Harvey, 1973, pp. 128–9)

These sentiments were echoed by Berry, also in 1973. The starting point for his plaint is that for any empirically observed spatial regularity two or more apparently plausible explanations can be offered. The investigator is then hard put to it to choose which of the various explanations is the most likely to be right; in many cases, this choice appears impossible to make. Both Berry and Harvey conclude that the time has come to seek a new paradigm, since the present one 'is ripe for overthrow' (Harvey, 1973, p. 129).

This impatience with the present situation seems to be predicated on two fundamental misconceptions. In the first place, Berry apparently supposes that because we cannot *now* choose between various explanations for phenomena it will never be possible for us so to do. But, as noted in Chapter 5 in the discussion of the urban rank–size relationship, the logic of an explanation can be explored to see whether it generates subsidiary hypotheses that can be tested. Thus, once the excitement of the frontier days is over, there is still a lot of hard and methodical work required to fill in the details and test the plausibility of explanations. If we refer back to Figure 8 (p. 52), the fact is that only a small part of the cube can so far be described by the ends of the vectors labelled 'good'; given the relatively short time for which social scientists in general have been working with the 'scientific' method, this is not altogether surprising.

The second misconception, and to my mind much the more serious one, is to suppose that because one paradigm (the 'scientific' paradigm) is inadequate to solve *all* our problems it is *altogether* useless. In the previous chapter I have argued that to produce a single theory of location is impossible; even to articulate the various partial theories is probably too much to expect. The distinction was also made between normative and positive theory. These points link back to Chapter 2 (p. 48), where even the generality of the 'scientific' method was questioned. Thus, while I am fully sympathetic to the view that the 'scientific' paradigm is not adequate to all our needs, and must be supplemented by other approaches, I am not persuaded that it should be replaced. And the interesting thing is that while Berry urges us to take a behavioural view of the world, which appears to mean the examination of individual perceptions and actions, Harvey wants us to embrace the Marxist 'method' of dialectic. This 'method' passes my understanding; so far as it has a value, it seems to be as a metaphysical belief system and not – as its protagonists proclaim – a mode of rational argument.

The limitations of positive theory

Rather than calling for a new paradigm to replace the scientific methodology with which we have been equipping ourselves, it is altogether more appropriate to examine the power and the limitations of the scientific method. In making that examination, two intertwined themes must be disentangled. The first has already been noted, namely, that once any kind of empirical pattern has been identified two or more rival theories may be available to explain its occurrence. To evaluate these rival theories can be a laborious business.

The second thread in the argument is much more important. Inferential statistics are designed for establishing and probing the relationships that exist between real-world phenomena. The analytical techniques are admirably designed to tease out the causal links in complex systems. Once the system has been taken apart, some reasonable inferences can be drawn regarding both the nature and the magnitude of changes likely to occur in the whole system if specified changes are made. For example, one might hope to predict the effect on house prices in a city arising from a given influx of new jobs and hence immigrants. Or, to take the converse case with contemporary significance, by how much ought wages for teachers, bus drivers, etc. in London be raised so that enough people will be attracted to the Metropolis to man these services effectively? However sophisticated this kind of positive analysis may become, it suffers from one crucial limitation. No amount of positive study will tell us how things *should* be organized: we cannot follow the positive road and hope to construct normative theory.

Normative theory ought to be based on the formal rules of logic and thus be internally consistent. It also ought to start from sensible assumptions, which can be checked by empirical observation. However, there is no proof that one set of assumptions is inherently better as the basis for postulating normative theory than is another. In the last resort, the choice between one set of basic assumptions and another is a metaphysical matter, a question of belief that is not amenable to scientific analysis. Or, as Morrill (1973, p. 5) put it: 'Desired transformations will not occur just because we study "relevant" problems and even find solutions ... geography must have a strong voice of conscience.'

The failure to realize that quantitative methods are generally positive in character rather than normative, and that choice between competing normative views of the world depends largely upon metaphysical matters of belief, probably underlies the disenchantment that some geographers have expressed. However, when these distinctions are made, the developments in recent decades can be seen in perspective as being extraordinarily useful but as yet neither exploited to the full nor as having their limitations specified. These are major tasks for the next decade and more.

The unfolding horizons

If the new wine of the 'scientific method' has in the past been put in old problem bottles, the search is now beginning for suitable new bottles. The most evident form of this search is the plea that geographical studies should be 'relevant', that is, relating to identifiable contemporary issues of importance (see for example *Area*, vols. 3, 4 and 5, 1971–3). This is a matter that bristles with difficulties, for who is to judge what is relevant to whom? Certainly, there is no consensus as to the extent to which geographical work should be 'relevant'. Nevertheless, there is widespread acceptance that a greater proportion of work could reasonably be directed to the solution of real and pressing economic and social problems. In this respect, the information contained in the table on page 178 is intriguing, summarizing the research directions identified in three recent publications. The reviews compiled by Chisholm and Taaffe were prepared for the Social Science Research Councils of the United Kingdom and United States respectively, and were intended to identify research areas where additional effort would be rewarding. Both of these lists may be regarded as essentially conservative, reflecting the evolutionary continuity of the subject. In striking contrast are the headings under which Albaum grouped the essays that he collected together. Here there is manifest a strong commitment to change society, not merely analyse it. Problems identified are essentially intra-urban, if urban is a term used with a wide definition, and concerned with the inequalities of life – inequalities related to income, ethnic differentiation, etc. – or indeed the general awfulness of life for many people. In this respect, Albaum has pointed an essentially new road, along which only a few intrepid explorers have so far ventured.

In contrast to Harvey's (1973) polemical essays on social justice in the city, D. M. Smith (1973a) in the same year systematically set about the task of relating the welfare aspects of spatial distributions to orthodox ideas in welfare economics. Whatever the given total of available resources, by what criteria can we say that one spatial arrangement is fairer or more just than another? If the structure of a city is changed, so altering the distribution of housing of various qualities, or if smokeless zones are enforced as a means for cleaning up the city's atmosphere, who gains most and will anyone be worse off? Since the wealthy have traditionally avoided urban atmospheric pollution by living above it (for example, Hampstead in London) or far enough away and especially on the windward side of cities, smoke abatement generally benefits most of those who live in or near the city centre, usually the poorest section of the community. Conversely,

EMERGENT FOCI FOR GEOGRAPHICAL WORK

United States Social Science Research Council (Taaffe)[1]	United Kingdom Social Science Research Council (Chisholm)[2]	M. Albaum[3]
1. Locational analysis, incl. behavioural studies and simulation	1. Perception studies	1. Geography and the transformation of society
2. Cultural geography, incl. cultural ecology and perception	2. Simulation models	2. Poverty and the poor in America
3. Urban studies, city systems and intra-urban problems	3. Forecasting	3. Black America and the black ghetto
4. Environmental and spatial behaviour	4. Regional taxonomy	4. Urban life and housing
	5. Environmental standards	5. Environmental deterioration, hazards and stress
	6. Population and migration	6. Population growth and pressure
	7. Processes of regional economic and social development	7. Conflict and conflict resolution

1. Taaffe, 1970
2. Chisholm, 1971b
3. Albaum, 1973

though, the establishment and extension of green belts around conurbations is of doubtful value to city-centre dwellers and probably most benefits the middle- to upper-income groups, who either live near the open spaces or can afford to travel to them. The greatest gain, however, may well be for those lucky enough to own a house in, or on the edge of, the green belt, as the capital value of the property is greatly enhanced by the statutory designation. Strongly symptomatic of this awakening interest among geographers, and reflecting the fact that these problems exist at all scales from the national down to the local scale, are recent books by D. M. Smith (1973b) and Bunge (1971). The former is an admirable survey of the spatial variation in the United States in social 'health' as measured by a whole range of indicators other than the basic and rather crude one of income differences. And while one may not agree with Bunge's views on the Fitzgerald community, also in the U.S.A., there is no burking his deep sense of outrage at what has happened – or not happened – there.

Apart from the contrast between Albaum's topics and the other two lists, the table on p. 178 carries another message. There is substantial agreement on the importance of the processes which are at work to shape the patterns that are manifest in space, whether these are the processes of perception, migration or economic development, etc. At the very least, this emphasis upon processes of all kinds is noticeably shifting the balance of geographical work; it is also taking geographers into entirely new areas of inquiry. To establish this point beyond any doubt, recollect the very recent surge of interest in perception studies (p. 159) and note just two other new kinds of inquiry. Urban growth and development is far too complex a process for all the causal links to be unravelled with our existing level of knowledge. On the other hand, it is essential to know something of the effects that will be induced by specific changes. Hence, the urgent need for models, albeit crude and imperfect, of the kind reviewed by Batty (1972) and being developed by Martin and March (1972) among others. The essential characteristic of this and kindred work is the attempt to construct robust models that will allow one to simulate urban conditions given assumed changes in certain parameters. For example, if a good relationship can be established between kind and density of residential housing areas on the one hand and the volume of commuting trips on the other, then predictions may be made regarding the investment in transport facilities implied by various patterns of residential expansion. This type of problem arises very commonly in the field of planning and it is with planners, engineers and others that geographers have been engaged in building urban simulation models.

The second area of burgeoning interest relates to the processes of economic and social growth and change. Economists and others have been busy since the last world war developing theories of economic development. With a few notable exceptions (for instance, Myrdal, 1957, and Hirschman, 1958), economists have treated economic development in highly aggregate terms which abstract the process from its spatial domain; space was largely ignored as a constraint on the development process. As Brookfield (1973) notes, it was not until about 1960 that geographers began to take a serious interest in the development process, but since then there has been a very considerable awakening and the volume of literature is now too vast to be adequately recognized here (see Brookfield, 1973; Grigg, 1973). Until very recently, however, geographers, in common with others, have viewed development as a diffusion process, the diffusion of 'western' culture and technology. Brookfield's paper marks a major turning point, for he calls in question this diffusionist view of development and asks us to understand what is going on in the Third World countries faced with incredibly rapid change. To do so properly and thoroughly, it is essential to question our own value-systems, so that we may the better see development from the viewpoint, not of the developer, but of those being 'developed'.

If, then, we take it as axiomatic that henceforth geographers will be numerate and that the subject has been re-orientated, away from description and synthesis toward the solution of problems, what are likely to be the characteristics of human geography in the next one or two decades? Although they are closely interrelated, it is convenient for the purpose of exposition to distinguish four themes.

During the nineteenth century and earlier, geography grew in response to contemporary needs, among other things for accurate maps and data about places. For perhaps the first half of the twentieth century, practitioners of the subject failed to notice that these needs are changing. The most dramatic change is associated with urbanization; in terms of population numbers, towns are rapidly becoming the single most important geographical phenomenon. Consequently, it is the problems of towns that ought to take pride of place in geographical studies, how they work and can/cannot be changed in a deliberate manner. Harvey's recent book *Social Justice and the City* is symptomatic of this concern, as is the related work by Pahl (1970, 1971) and the attempts to model urban systems mentioned above. The field is vast, including questions of intra-urban migration, the way the social class of residential areas changes upwards and downwards as well as the whole matter of access to urban resources.

Associated with urbanization is pollution and acute pressure on resources of all kinds, whether consumable minerals or recreational open space. The world is waking up to the fact that ecological limits are being reached or at least seem perilously close at hand. Geographers are now remembering their traditional concern for man/environment issues and in particular are examining the ways in which ecological systems actually work (see, for example, Manners and Mikesell, 1974).

Common to both of these trends is a third emergent characteristic of geography, namely, an emphasis upon practical applications to solve actual problems. One of the more striking examples of this development is the fact that among graduate entrants to the planning profession in Britain, geographers form the largest group of all the disciplines. In this respect, geography is shifting from the stance of seeking to explain phenomena to the position of actually doing something about the matter at hand. The implication is the need to acquire practical skills to a level which will enable the student to use them. In this respect, the subject is becoming altogether more 'professional'.

Taken together, these three trends appear to be paving the way for a fourth. As noted above the new quantitative techniques have hitherto been applied to fairly traditional geographical problems but at a level below that of the overriding question of man's relationship to the natural environment. The problems involved have appeared too formidable and, in the disenchantment with environmentalism, few have ventured into this arena. However, with the recognition of the pressing problems that face the world and with the development of scientific techniques in general, the time is now opportune for geographers to pick up the ecological threads that ramify through the history of the subject but which have lain largely neglected in recent decades (Morgan and Moss, 1967). Substantive contributions have been made by several authors, notably Sewell (1969, 1970) in the field of resource management in general and of water in particular.

Thus, in this brief journey into the history of geography, we have travelled through two major phases and have embarked upon a third one. The first of these periods was concerned with the description of static patterns, the accurate recording of information in a spatial domain. The second in part initiated the era of modern quantitative methods, through attempts to establish relationships between these static distributions. Now we are launched upon the third and current phase, which is essentially to sort out the nature of the operative processes. For this purpose, dynamic patterns, or processes of change (Chapter 4) must be related to explanatory theory (Chapter 5). As long ago as 1953, Schaefer wrote: 'mature social science looks for process laws'. That is what geography is now about, in an attempt to forge coherent links between spatial forms and spatial processes. The focus of attention is concurrently shifting from rural to essentially urban problems and the purpose is to offer, if not answers to pressing social and economic problems, at least means whereby such answers may be sought. No longer content to explain what is, geographers are seeking to contribute directly to what might be. Perhaps *the* great change in the subject called geography is the awareness of the future to be moulded as against the past to be explained. It is this fundamental shift in outlook which above all others colours the work of the present generation, making today's geography so different from yesterday's.

References

Abler, R., J. S. Adams and P. Gould (1971), *Spatial Organization. The Geographer's View of the World*, Prentice-Hall.

Albaum, M. (ed.) (1973), *Geography and Contemporary Issues: Studies of Relevant Problems*, Wiley.

Allen, G. R. (1954), 'The "courbe des populations", a Further Analysis', *Bulletin of the Oxford Institute of Statistics*, 16, 179–89.

Allen, R. G. D. (1966), *Statistics for Economists*, 3rd edn, Hutchinson.

Andrews, H. F. (1970), 'Working Notes and Bibliography on Central Place Studies', *Discussion Paper* 8, Department of Geography, University of Toronto.

Angel, S., and G. Hyman (1971), *Transformations and Geographic Theory*, Centre for Environmental Studies.

Aron, R. (1965), *Main Currents in Sociological Thought I*, Penguin.

Bagnold, R. A. (1941), *The Physics of Blown Sand and Desert Dunes*, Methuen.

Baker, A. R. H. (1972), *Progress in Historical Geography*, David and Charles.

Baker, J. N. L. (1945), *A History of Geographical Discovery and Exploration*, Harrap.

Bassett, K. (1972), 'Numerical Methods for Map Analysis', in C. Board et al. (eds.), *Progress in Geography*, Arnold, 4, 217–54.

Bassett, K. A., and G. B. Norcliffe (1969), 'Filter Theory and Filter Methods in Geographical Research', *Seminar Paper Series*, Series A, Number 20, Department of Geography, University of Bristol.

Batty, M. (1972), 'Recent Developments in Land-Use Modelling: A Review of Current Research', *Urban Studies*, 9, 151–77.

Beckerman, W. (1956), 'Distance the Pattern of Intra-European Trade', *Review of Economics and Statistics*, XXXVIII, 31–40.

Beckmann, M. J. (1957–8), 'City Hierarchies and the Distribution of City Size', *Economic Development and Cultural Change*, 6, 243–8.

Berry, B. J. L. (1973), 'A Paradigm for Modern Geography', in R. J. Chorley (ed.), 3–21.

(1972), 'Hierarchical Diffusion: The Basis of Developmental Filtering and Spread in a System of Growth Centers', in N. M. Hansen (ed.), *Growth Centers in Regional Economic Development*, Free Press, 108–38.

(1967), *Geography of Market Centers and Retail Distribution*, Prentice-Hall.

(1966), 'Essays on Commodity Flows and the Spatial Structure of the Indian Economy', *Research Paper* 111, University of Chicago, Department of Geography.

(1964), 'Approaches to Regional Analysis: A Synthesis', *Annals*, Association of American Geographers, 54, 2–11. Reprinted in B. J. L. Berry and D. Marble (eds.), *Spatial Analysis. A Reader in Statistical Geography*, Prentice-Hall, 1968, 24–34.

(1961), 'A Method for Deriving Multifactor Uniform Regions', *Przeglad Geograficzny*, 33, 263–82.

Berry, B. J. L., and A. Pred (1961 and 1965), *Central Place Studies: A Bibliography of Theory and Applications*, Regional Science Research Institute.

Beveridge, W. S. (1921), 'Weather and Harvest Cycles', *Economic Journal*, 31, 429–52.

Bishop, K. C., and C. E. Simpson (1972), 'Components of Change Analysis: Problems of Alternative Approaches to Industrial Structure', *Regional Studies*, 6, 59–68.

Blache, P. V. de la (1903), *Géographie de la France*, Paris.

Blaut, J. M. (1961), 'Space and Process', *Professional Geographer*, 13. Reprinted in W. K. D. Davies, *The Conceptual Revolution in Geography*, University of London Press, 1972, 42–51.

Boulding, K. E. (1970), *Economics as a Science*, McGraw-Hill.

Bromhead, P. (1973), *The Great White Elephant of Maplin Sands*, Paul Elek.

Brooke, M. Z. (1970), *Le Play. Engineer and Social Scientist*, Longmans.

Brookfield, H. C. (1973), 'On One Geography and a Third World', *Transactions*, Institute of British Geographers, 58, 1–20.

(1971), *Melanesia: A Geographical Interpretation of an Island World*, Methuen.

(1969), 'On the Environment as Perceived', in C. Board et al. (eds.), *Progress in Geography*, Arnold, vol. 1, 51–80.

Brown, E. H. Phelps (1972), 'The Underdevelopment of Economics', *Economic Journal*, 82, 1–10.

Brown, L. A. (1968a), 'Diffusion Dynamics. A Review and Revision of the Quantitative Theory of the Spatial Diffusion of Innovation', *Lund Studies in Geography*, Ser. B, 29.

(1968b), *Diffusion Processes and Location. A Conceptual Framework and Bibliography*, Regional Science Research Institute.

Brown, L. A., and K. R. Cox (1971), 'Empirical Regularities in the Diffusion of Innovation', *Annals*, Association of American Geographers, 61, 551–9.

Brown, L. A., and E. G. Moore (1969), 'Diffusion Research in Geography: A Perspective', in C. Board et al. (eds.), *Progress in Geography*, Arnold, vol. 1, 119–57.

Buckley, W. (ed.) (1968), *Modern Systems Research for the Behavioral Scientist*, Aldine.

Bunge, W. (1971), *Geography of a Revolution*, Schenkman.

(1966), 'Theoretical Geography', 2nd edn, *Lund Studies in Geography*, Ser. C, Gleerup.

Burton, I. (1963), 'The Quantitative Revolution and Theoretical Geography', *The Canadian Geographer*, 7, 151–62. Reprinted in B. J. L. Berry and D. F. Marble (eds.), *Spatial Analysis. A Reader in Statistical Geography*, Prentice-Hall, 1968.

Carson, M. A., and M. J. Kirkby (1972), *Hillslope Form and Process*, C.U.P.

Chalmers, J. A. (1971), 'Measuring Changes in Regional Industrial Structure: A Comment on Stilwell and Ashby', *Urban Studies*, 8, 289–92.

Champernowne, D. G. (1973), Review in the *Economic Journal*, 83, 908–10.

Chisholm, M. (1971a), 'In Search of a Basis for Location Theory: Micro-Economics or Welfare Economics?', in C. Board et al. (eds.), *Progress in Geography*, Arnold, vol. 3, 111–33.

(1971b), *Research in Human Geography*, Heinemann, for the Social Science Research Council.

(1971c), 'Freight Transport Costs, Industrial Location and Regional Development', in M. Chisholm and G. Manners (eds.), 213–44.

(1966), *Geography and Economics*, Bell.

(1964), 'Problems in the Classification and Use of Farming-Type Regions', *Transactions and Papers*, Institute of British Geographers, 35, 91–103.

Chisholm, M. – *contd*

(1962), *Rural Settlement and Land Use: An Essay in Location*, Hutchinson.

Chisholm, M., and P. O'Sullivan (1973), *Freight Flows and Spatial Aspects of the British Economy*, C.U.P.

Chisholm, M., and J. Oeppen (1973), *The Changing Pattern of Employment. Regional Specialisation and Industrial Localisation in Britain*, Croom Helm.

Chisholm, M., and B. Rodgers (eds.) (1973), *Studies in Human Geography*, Heinemann.

Chisholm, M., and G. Manners (eds.) (1971), *Spatial Policy Problems of the British Economy*, C.U.P.

Chorley, R. J. (ed.) (1973), *Directions in Geography*, Methuen.

(1970), 'The Application of Quantitative Methods to Geomorphology', in R. J. Chorley and P. Haggett (eds.), *Frontiers in Geographical Teaching*, 2nd edn, Methuen, 147–63.

Chorley, R. J., and B. A. Kennedy (1971), *Physical Geography: A Systems Approach*, Prentice-Hall.

Chorley, R. J., and P. Haggett (eds.) (1967), *Models in Geography*, Methuen.

(1965), 'Trend-Surface Mapping in Geographical Research', *Transactions*, Institute of British Geographers, 37, 47–67. Reprinted in B. J. L. Berry and D. F. Marble, *Spatial Analysis*, Prentice-Hall, 1968, 195–217.

Christaller, W. (1933 and 1966), *Central Places in Southern Germany*, translated by C. W. Baskin, Prentice-Hall. First published in German.

Clark, C. (1951), 'Urban Population Densities', *Journal*, Royal Statistical Society, Series A, 114, 490–96.

Clark, C., F. Wilson and J. Bradley (1969), 'Industrial Location and Economic Potential in Western Europe', *Regional Studies*, 3, 197–212.

Clark, D. (1973), 'The Formal and Functional Structure of Wales', *Annals*, Association of American Geographers, 63, 71–84.

Cliff, A. D. (1970), 'Computing the Spatial Correspondence between Geographical Patterns', *Transactions*, Institute of British Geographers, 50, 143–54.

(undated), 'Some Measures of Contiguity for Two Colour Mosaic Maps with Vacancies', *Seminar Paper Series*, Ser. A, No. 4, Department of Geography, University of Bristol.

Cliff, A. D., and J. K. Ord (1969), 'The Problem of Spatial Autocorrelation', in A. J. Scott (ed.), *Studies in Regional Science*, Pion, 25–55.

Coppock, J. T. (1971), *An Agricultural Geography of Great Britain*, Bell.

(1964), 'Crop, Livestock, and Enterprise Combinations in England and Wales', *Economic Geography*, 40, 65–81.

Cordey-Hayes, M. (1968), 'Retail Location Models', *Working Paper* 16, Centre for Environmental Studies.

Crone, G. R. (1968), *Maps and their Makers*, 4th edn, Hutchinson.

Curry, L. (1966), 'Chance and the Landscape', in J. W. House (ed.), *Northern Geographical Essays in Honour of G. H. J. Daysh*, Oriel Press, 40–55.

(1964), 'The Random Spatial Economy: An Exploration in Settlement Theory', *Annals*, Association of American Geographers, 54, 138–46.

Dacey, M. F. (1973), 'Some Questions about Spatial Distributions', in R. J. Chorley (ed.) (1973), 127–51.

(1965), *A Review on Measures of Contiguity for Two and K-Color Maps*, Technical Report No. 2, Spatial Diffusion Study, Department of Geography, Northwestern University.

Darby, H. C. et al. (1952–), *The Domesday Geography of England*, C.U.P.

(1940), *The Draining of the Fens*, C.U.P.

(ed.) (1936), *An Historical Geography of England before A.D. 1800*, C.U.P.

Davies, R. (1973), 'The Location of Service Activities', in M. Chisholm and B. Rodgers (eds.) (1973), 125–71.

Davis, W. M. (1909), *Geographical Essays*, first published in 1909 and re-issued by Dover, 1954.

Demolins, E. (1901–3), *Comment la route crée le type social*, 2 vols., Firmin-Didot.

Dickinson, G. C. (1963), *Statistical Mapping and the Presentation of Statistics*, Arnold.

Dickinson, R. E. (1969), *The Makers of Modern Geography*, Routledge and Kegan Paul.

(1947), *City, Region and Regionalism: A Geographical Contribution to Human Ecology*, Kegan Paul.

Downs, R. M. (1970), 'Geographic Space Perception: Past Approaches and Future Prospects', in C. Board et al. (eds.), *Progress in Geography*, Arnold, vol. 2, 65–108.

Dury, G. (1970), 'Merely from Nervousness', *Area*, 2, 29–32.

East, W. G., O. H. K. Spate and C. A. Fisher (1971), *The Changing Map of Asia. A Political Geography*, 5th revised edn, Methuen.

Estall, R. C., and R. O. Buchanan (1961 and 1966), *Industrial Activity and Economic Geography*, Hutchinson, revised edn 1966.

Febvre, L. (1925), *Geographical Introduction to History*. English translation by E. G. Mountford, Knopf.
Ferguson, A. G., and P. C. Forer (1973), 'Aspects of Measuring Employment Specialization in Great Britain', *Area*, 5, 121–8.
Fisher, C. A. (1970), 'Whither Regional Geography?', *Geography*, LV, 373–89.
Florence, P. S. (1944), 'The Selection of Industries Suitable for Dispersion into Rural Areas', *Journal*, Royal Statistical Society, CVII, 93–107.
Forster, F. (1966), 'Use of Demographic Base Maps for the Presentation of Areal Data in Epidemiology', *British Journal of Preventive and Social Medicine*, 20, 165–71.
Freeman, T. W. (1967), *The Geographer's Craft*, Manchester University Press.
(1961), *A Hundred Years of Geography*, Duckworth.
Friedlander, D., and D. J. Roshier (1966), 'A Study of Internal Migration in England and Wales, Part I. Geographical Pattern of Internal Migration 1851–1951', *Population Studies*, XIX, 239–79.
Garner, B. (1967), 'Models of Urban Geography and Settlement Location', in R. J. Chorley and P. Haggett (eds.) (1967), 303–60.
Garrison, W. L. (1960), 'Connectivity of the Interstate Highway System', *Papers and Proceedings*, Regional Science Association, 6, 121–37.
Goddard, J. B. (1973), 'Office Linkages and Location. A Study of Communications and Spatial Patterns in Central London', *Progress in Planning*, 1:2, edited by D. Diamond and J. B. McLoughlin, Pergamon.
Goheen, P. G. (1970), *Victorian Toronto, 1850 to 1900: Pattern and Process of Growth*, University of Chicago, Department of Geography, Research Paper 127.
Gordon, I. R. (1973), 'The Return of Regional Multipliers: A Comment', *Regional Studies*, 7, 257–62.
Gottmann, J. (1961), *Megalopolis: The Urbanized Northeastern Seaboard of the United States*, M.I.T. Press.
Goudie, A. S., B. Allchin and K. T. M. Hegde (1973), 'The Former Extensions of the Great Indian Sand Desert', *Geographical Journal*, 139, 243–57.
Gould, P. R. (1969a), 'Methodological Developments since the Fifties', in C. Board et al. (eds.), *Progress in Geography*, Arnold, vol. 1, 1–49.
(1969b), 'Spatial Diffusion', *Resource Paper* 4, Commission on College Geography, Association of American Geographers.

(1963), 'Man against His Environment: A Game Theoretic Framework', *Annals*, Association of American Geographers, 53, 290–97.

Gould, P. R., and R. R. White (1968), 'The Mental Maps of British School Leavers', *Regional Studies*, 2, 161–82.

Granger, C. W. J. (1969), 'Spatial Data and Time Series Analysis', in A. J. Scott (ed.) (1969), 1–24.

Grant, F. (1957), 'A Problem in the Analysis of Geographical Data', *Geophysics*, 22, 309–44.

Gregory, S. (1963), *Statistical Methods and the Geographer*, Longmans.

Grigg, D. (1973), 'Geographical Studies of Economic Development with Special Reference to Agriculture', in M. Chisholm and B. Rodgers (eds.) (1973), 18–84.

Grove, A. T., and A. Warren (1968), 'Quaternary Landforms and Climate on the South Side of the Sahara', *Geographical Journal*, 134, 194–208.

Guelke, L. (1971), 'Problems of Scientific Explanation in Geography', *Canadian Geographer*, xv, 38–53.

Hägerstrand, T. (1953 and 1967), *Innovation Diffusion as a Spatial Process*, University of Chicago Press. First published in Swedish in 1953. Translated by A. Pred, 1967.

Haggett, P. (1972), *Geography: A Modern Synthesis*, Harper & Row.

(1965), *Locational Analysis in Human Geography*, Arnold.

(1964), 'Regional and Local Components in the Distribution of Forested Areas in South East Brazil: A Multivariate Approach', *Geographical Journal*, 130, 365–78.

Haggett, P., and R. J. Chorley (1969), *Network Analysis in Geography*, Arnold.

Hall, P. (ed.) (1966), *Von Thünen's Isolated State*, Pergamon.

Hall, P. et al. (1973), *The Containment of Urban England*, 2 vols., Allen and Unwin.

Hare, F. K. (1951), 'Climatic Classification', in L. D. Stamp and S. W. Wooldridge (eds.), *London Essays in Geography. Rodwell Jones Memorial Volume*, Longmans Green, 111–34.

Häro, A. S. (1968), 'Area Cartogram of the SMSA Population of the United States', *Annals*, Association of American Geographers, 58, 452–60.

Harris, C. (1971), 'Theory and Synthesis in Historical Geography', *Canadian Geographer*, xv, 157–72.

Harris, C. D. (1954), 'The Market as a Factor in the Localization of Industry in the United States', *Annals*, Association of American

Geographers, 44, 315–48. Reprinted in R. H. T. Smith, E. J. Taaffe and L. J. King (eds.), *Readings in Economic Geography; The Location of Economic Activity*, Rand McNally, 1968.
Hart, R. A. (1970), 'A Model of Inter-Regional Migration in England and Wales', *Regional Studies*, 4, 279–96.
Hartshorne, R. (1960), *Perspective on the Nature of Geography*, John Murray.
(1939), *The Nature of Geography*, Association of American Geographers.
Hartshorne, R., and S. N. Dicken (1935), 'A Classification of the Agricultural Regions of Europe and North America on a Uniform Statistical Basis', *Annals*, Association of American Geographers, 25, 99–120.
Harvey, D. (1973), *Social Justice and the City*, Arnold.
(1969), *Explanation in Geography*, Arnold.
(1968), 'Pattern, Process, and the Scale Problem in Geographical Research', *Transactions*, Institute of British Geographers, 45, 71–8.
(1967), 'The Problem of Theory Construction in Geography', *Journal of Regional Science*, 7, 211–16.
Hay, A. (1973), *Transport for the Space Economy. A Geographical Study*, Macmillan.
Hay, A. M., and R. H. T. Smith (1970), *Interregional Trade and Money Flows in Nigeria, 1964*, O.U.P.
Herbertson, A. J. (1905), 'The Major Natural Regions: An Essay in Systematic Geography', *Geographical Journal*, 25, 300–310.
Hettner, A. (1921 and 1972), *The Surface Features of the Land*, Macmillan. First published in German in 1921 and translated by P. Tilley.
Hill, C. P. (1972), *British Economic and Social History 1700–1964*, 3rd edn, Arnold.
Hirschman, A. O. (1958), *The Strategy of Economic Development*, Yale University Press.
Hirst, M. A. (1973), 'Administrative Reorganization in Uganda: Towards a More Efficient Solution', *Area*, 5, 177–81.
Hoag, L. P. (1969), 'The Weaver Method: An Evaluation', *Professional Geographer*, XXI, 244–6.
Hodder, B. W., and U. I. Ukwu (1969), *Markets in West Africa. Studies of Markets and Trade Among the Yoruba and Ibo*, Ibadan University Press.
Hodder, I. R. (1972a), 'The Interpretation of Spatial Patterns in Archaeology: Two Examples', *Area*, 4, 223–9.
(1972b), 'Locational Models and the Study of Romano-British Settle-

ment', in D. L. Clarke (ed.), *Models in Archaeology*, Methuen, 887–909.

Hollingsworth, T. H. (1965), page 132 of the Reader's Digest *Complete Atlas of the British Isles*. First published in *The Times*, 19 October 1964.

Hoover, E. M. (1948), *The Location of Economic Activity*, McGraw-Hill.

(1936), 'The Measurement of Industrial Localization', *Review of Economic Statistics*, 18, 162–71.

Horton, R. E. (1945), 'Erosional Development of Streams and Their Drainage Basins: Hydrophysical Approach to Quantitative Morphology', *Bulletin*, Geological Society of America, LVI, 275–370. Reprinted in G. H. Dury (ed.), *Rivers and Terraces*, Macmillan (1970), 117–65.

Hotelling, H. (1929), 'Stability in Competition', *Economic Journal*, 39, 41–57.

Hudson, J. C. (1969), 'Diffusion in a Central Place System', *Geographical Analysis*, 1, 45–58.

Huff, D. L. (1973), 'The Delineation of a National System of Planning Regions on the Basis of Urban Spheres of Influence', *Regional Studies*, 7, 323–9.

Huntington, E. (1915), *Civilization and Climate*, Yale University Press.

(1907), *The Pulse of Asia. A Journey in Central Asia Illustrating the Geographic Basis of History*, Houghton Mifflin.

Isard, W. (1956), *Location and Space-Economy. A General Theory Relating to Industrial Location, Market Areas, Land Use, Trade, and Urban Structure*, M.I.T. Press and Wiley.

Isard, W. et al. (1969), *General Theory. Social, Political, Economic, and Regional, with Particular Reference to Decision-Making Analysis*, M.I.T. Press.

James, P. E. (1971), *On Geography. Selected Writings of Preston E. James*, edited by D. W. Meinig, Syracuse University Press.

Janelle, D. G. (1968), 'Central Place Development in a Time-Space Framework', *Professional Geographer*, XX, 5–10.

Jevons, W. S. (1909), *Investigations in Currency and Finance*, edited by H. S. Foxwell, Macmillan.

Johnson, L. J., and W. E. Teufner (1968), 'Industry Combinations in the Central States: An Application of Weaver's Method', *Professional Geographer*, XX, 297–302.

Johnston, R. J. (1973), *Spatial Structures. Introducing the Study of Spatial Systems in Human Geography*, Methuen.

Jonasson, O. (1925), 'Agricultural Regions of Europe', *Economic Geography*, 1, 277–314.

Kaldor, N. (1972), 'The Irrelevance of Equilibrium Economics', *Economic Journal*, 82, 1,237–55.

Kansky, K. J. (1963), 'Structure of Transportation Networks: Relationships between Network Geometry and Regional Characteristics', *Research Paper* 84, University of Chicago, Department of Geography.

Keeble, D. (1971), 'Employment Mobility in Britain', in M. Chisholm and G. Manners (eds.) (1971), 24–68.

Kimble, G. H. T. (1951), 'The Inadequacy of the Regional Concept', in L. D. Stamp and S. W. Wooldridge (eds.), *London Essays in Geography. Rodwell Jones Memorial Volume*, Longmans Green, 151–74.

King, L. J. (1969a), *Statistical Analysis in Geography*, Prentice-Hall.

(1969b), 'The Analysis of Spatial Form and Its Relation to Geographic Theory', *Annals*, Association of American Geographers, 59, 573–95.

König, D. (1936 and 1950), *Theorie der endlichen und unendlichen Graphen: kombinatorische Topologie der Streckenkomplexe*, 1950 edn published by Chelsea Publications.

Krumbein, W. C. (1956), 'Regional and Local Components in Facies Maps', *Bulletin of the American Association of Petroleum Geology*, 40, 2,163–94.

Lambert, A. M. (1971), *The Making of the Dutch Landscape*, Seminar Press.

Lanchester Polytechnic (1969), *Gravity Models in Town Planning*, mimeo.

Land, A. H. (1957), 'An Application of Linear Programming to the Transport of Coking Coal', *Journal*, Royal Statistical Society, 120, 308–19.

Langton, J. (1972), 'Potentialities and Problems of Adopting a Systems Approach to the Study of Change in Human Geography', in C. Board et al. (eds.), *Progress in Geography*, vol. 4, Arnold, 125–79.

Lawton, R. (1968), 'The Journey to Work in Britain: Some Trends and Problems', *Regional Studies*, 2, 27–40.

(1963), 'The Journey to Work in England and Wales: Forty Years of Change', *Tijdschrift voor Economische en Sociale Geografie*, 54, 61–9.

Leopold, L. B., M. G. Wolman and J. P. Miller (1964), *Fluvial Processes in Geomorphology*, Freeman.

Lever, W. F. (1973), 'A Markov Approach to the Optimal Size of Cities in England and Wales', *Urban Studies*, 10, 353–65.

Lewis, W. V. (ed.) (1960), *Norwegian Cirque Glaciers*, Royal Geographical Society Research Series, No. 4.

(1940), 'Dirt Cones on the Northern Margins of Vatnjökull, Iceland', *Journal of Geomorphology*, 3, 16–26.

Liepmann, K. K. (1944), *The Journey to Work. Its Significance for Industrial and Community Life*, Kegan Paul.

Linneman, H. (1966), *An Econometric Study of International Trade Flows*, North Holland.

Lösch, A. (1954), *The Economics of Location*, Yale University Press. Translated from the German by W. F. Stolper.

McCarty, H. H. (1954), 'An Approach to a Theory of Economic Geography', *Economic Geography*, 30, 95–101.

McCarty, H. H., J. C. Hook and D. S. Knos (1956), *The Measurement of Association in Industrial Geography*, State University of Iowa, Department of Geography.

McLoughlin, J. B. (1969), *Urban and Regional Planning. A Systems Approach*, Faber.

Macmillan, W. (1973), 'The Entropy Maximising Approach to the Journey to Work Problem', mimeo, Department of Geography, University of Bristol.

Manley, G. (1958), 'The Revival of Climatic Determinism', *Geographical Review*, 48, 98–105.

Manners, I. R., and M. W. Mikesell (eds.) (1974), *Perspectives on Environment*, Association of American Geographers.

Markham, S. F. (1942), *Climate and the Energy of Nations*, O.U.P.

Martin, L., and L. March (eds.) (1972), *Urban Space and Structures*, C.U.P.

Masser, I. (1970), 'A Test of Some Models for Predicting Intermetropolitan Movement of Population in England and Wales', *University Working Paper* 9, Centre for Environmental Studies.

Massey, D. (1968), 'Problems of Location: Game Theory and Gaming Simulation', *Working Papers* 15, Centre for Environmental Studies.

May, J. A. (1970), *Kant's Concept of Geography and Its Relation to Recent Geographical Thought*, University of Toronto Press.

Miller, R. L. (1956), 'Trend Surfaces: Their Application to Analysis and Description of Environments of Sedimentation', *Journal of Geology*, 64, 425–46.

Mills, E. S. (1972), *Urban Economics*, Scott, Foresman.

(1970), 'Urban Density Functions', *Urban Studies*, 7, 5–20.

Mills, G. (1969), *Introduction to Linear Algebra for Social Scientists*, Allen and Unwin.

Minshull, R. C. (1967), *Regional Geography: Theory and Practice*, Hutchinson.

Monkhouse, F. J., and H. R. Wilkinson (1952), *Maps and Diagrams: Their Compilation and Construction*, Methuen, 3rd edn, 1971.

Morgan, W. B., and R. P. Moss (1967), 'Geography and Ecology: The Concept of the Community and Its Relationship to Environment', *Annals*, Association of American Geographers, 55, 339–50.

Morisawa, M. (1968), *Streams: Their Dynamics and Morphology*, McGraw-Hill.

Morrill, R. L. (1973), 'Geography and the Transformation of Society', in M. Albaum (ed.) (1973), 1–8.

(1971), 'On the Arrangement and Concentration of Points in the Plane', in H. McConnell and D. W. Yaseen (eds.), *Perspectives in Geography I. Models of Spatial Variation*, Northern Illinois University Press, 29–43.

(1970), 'The Shape of Diffusion in Space and Time', *Economic Geography*, 46 (supplement), 259–68.

(1965), 'Migration and the Spread and Growth of Urban Settlement', *Lund Studies in Geography*, Ser. B, Human Geography, No. 26.

Myrdal, G. (1957), *Economic Theory and Underdeveloped Regions*, Duckworth.

Newman, J. L. (1973), 'The Use of the Term "Hypothesis" in Geography', *Annals*, Association of American Geographers, 63, 22–7.

Norcliffe, G. (1969), 'On the Use and Limitations of Trend Surface Models', *Canadian Geographer*, XIII, 338–48.

Nystuen, J. D., and M. F. Dacey (1961), 'A Graph Theory Interpretation of Nodal Regions', *Papers and Proceedings*, Regional Science Association, 7, 29–42.

Ohlin, B. (1933), *Interregional and International Trade*, Harvard University Press. Revised edn, 1967.

Olsson, G. (1965), *Distance and Human Interaction: A Review and Bibliography*, Regional Science Research Institute.

Pahl, R. E. (1971), 'Poverty and the Urban System', in M. Chisholm and G. Manners (eds.), 126–45.

(1970), *Patterns of Urban Life*, Longmans.

Passarge, S. (1929), *Beschreibende Landschaftskunde*, Hamburg.

Paterson, J. H. (1960), *North America. A Regional Geography*, O.U.P.

Penck, W. (1925), *Morphological Analysis of Land Forms*, first published in German. Translated by H. Czech and K. C. Boswell and published by Macmillan in 1953.

Pitty, A. F. (1971), *Introduction to Geomorphology*, Methuen.

Planhol, X. de (1972), 'Historical Geography in France', in A. R. H. Baker (ed.) (1972), 29–44.

Pounds, N. J. G. (1973), *An Historical Geography of Europe 450 BC–AD 1330*, C.U.P.

Pred, A. R. (1973), 'Urbanisation, Domestic Planning Problems and Swedish Geographic Research', in C. Board et al., *Progress in Geography*, vol. 5, Arnold, 1–76.

(1967 and 1969), 'Behavior and Location: Foundations for a Geographic and Dynamic Location Theory', in two parts, *Lund Studies in Geography*, Ser. B, 27 and 28.

(1966), *The Spatial Dynamics of U.S. Urban-Industrial Growth, 1800–1914: Interpretative and Theoretical Essays*, M.I.T. Press.

Pred, A. R., and G. E. Törnqvist (1973), 'Systems of Cities and Information Flows', *Lund Studies in Geography*, Ser. B, 38.

Raisz, E. (1938), *General Cartography*, McGraw-Hill.

Ravenstein, E. G. (1885), 'The Laws of Migration', *Journal*, Royal Statistical Society, 48, 167–235.

Rawstron, E. M. (1958), 'Three Principles of Industrial Location', *Transactions and Papers*, Institute of British Geographers, 25, 135–42.

Reclus, E. (1876–87), *Nouvelle Géographie Universelle*, Hachette, 12 vols.

Reilly, W. J. (1929), *Methods for the Study of Retail Relationships*, Bureau of Business Research, University of Texas.

Richardson, H. W. (1973a), *The Economics of Urban Size*, Saxon House.

(1973b), 'Theory of the Distribution of City Sizes: Review and Prospects', *Regional Studies*, 7, 239–51.

Robinson, A. H., and R. A. Bryson (1957), 'A Method for Describing Quantitatively the Correspondence of Geographical Distributions', *Annals*, Association of American Geographers, 47, 379–91.

Robson, B. T. (1973a), 'A View on the Urban Scene', in M. Chisholm and B. Rodgers (eds.), 203–41.

(1973b), *Urban Growth. An Approach*, Methuen.

Rogers, A. (1971), *Matrix Methods in Urban and Regional Analysis*, Holden-Day.

Roxby, P. M. (1926), 'The Theory of Natural Regions', *Geographical Teacher* (*Geography*), XIII, 376–82.

Sack, R. D. (1972), 'Geography, Geometry and Explanation', *Annals*, Association of American Geographers, 62, 61–78.

Sauer, C. O. (1952), *Agricultural Origins and Dispersals*, American Geographical Society.

Schaefer, F. K. (1953), 'Exceptionalism in Geography: A Methodological Examination', *Annals*, Association of American Geographers, 43, 226–49.

Schwind, P. J. (1971), 'Migration and Regional Development in the United States, 1950–1960', *Research Paper*, 133, University of Chicago, Department of Geography.

Scott, A. J. (ed.) (1969), *Studies in Regional Science*, Pion.

Semple, E. C. (1911), *Influences of Geographic Environment*, Constable.

Sewell, W. R. D., and J. Rostron (1970), *Recreational Fishing Evaluation*, Department of Fisheries and Forestry, Canada.

Sewell, W. R. D., R. W. Judy and L. Ouellet (1969), *Water Management Research: Social Science Priorities*, Department of Energy, Mines and Resources, Canada.

Shonfield, A. (1972), 'The Social Sciences in the Great Debate on Science Policy', *Minerva*, x, 426–38.

Simon, H. A. (1957), *Models of Man: Social and Rational*, Wiley.

Singer, H. W. (1936), 'The "courbe des populations". A Parallel to Pareto's Law', *Economic Journal*, 46, 254–63.

Sinnhuber, K. A. (1954), 'Central Europe – Mitteleuropa – Europe Centrale: An Analysis of a Geographical Term', *Transactions and Papers*, Institute of British Geographers, 20, 15–39. The key map is reproduced in Haggett (1972), 85.

Skelton, R. A. (ed.) (1964), *History of Cartography*, Harvard University Press.

Smith, C. T. (1967), *An Historical Geography of Western Europe before 1800*, Longmans.

Smith, D. M. (1973a), 'An Introduction to Welfare Geography', *Occasional Paper* 11, Department of Geography, University of Witwatersrand.

(1973b), *The Geography of Social Well-Being in the United States. An Introduction to Territorial Social Indicators*, McGraw-Hill.

Spate, O. H. K. (1952), 'Toynbee and Huntington: A Study in Determinism', *Geographical Journal*, CXVIII, 406–24.

Spence, N. A. (1968), 'A Multifactor Uniform Regionalization of British Counties on the Basis of Employment Data for 1961', *Regional Studies*, 2, 87–104.

Steers, J. A. (ed.) (1934), *Scolt Head Island. The Story of Its Origin: The Plant and Animal Life of Its Dunes and Marshes*, Heffer.

Stevens, A. (1939), 'The Natural Geographical Region', *Scottish Geographical Magazine*, 55, 305–17.

Stewart, J. Q. (1947), 'Empirical Mathematical Rules Concerning the Distribution and Equilibrium of Population', *Geographical Review*, 37, 461–85.

Stewart, J. Q., and W. Warntz (1958), 'Physics of Population Distribution', *Journal of Regional Science*, 1, 99–123. Reprinted in B. J. L. Berry and D. F. Marble (eds.), *Spatial Analysis. A Reader in Statistical Geography*, Prentice-Hall (1968), 130–46.

Stoddart, D. R. (1967), 'Growth and Structure of Geography', *Transactions*, Institute of British Geographers, 41, 1–19.

(1966), 'Darwin's Impact on Geography', *Annals*, Association of American Geographers, 56, 683–98. Reprinted in W. K. D. Davies (ed.), *The Conceptual Revolution in Geography*, University of London Press (1972), 52–76.

(1965), 'Geography and the Ecological Approach. The Ecosystem as a Geographic Principle and Method', *Geography*, L, 242–51.

Strahler, A. N. (1950), 'Equilibrium Theory of Erosional Slopes Approached by Frequency Distribution Analysis', *American Journal of Science*, 248, 673–96 and 800–14.

Sylvester, D. (1969), *The Rural Landscape of the Welsh Borderland. A Study in Historical Geography*, Macmillan.

Szumeluk, K. (1968), 'Central Place Theory. I. A Review', *Working Papers* 2, Centre for Environmental Studies.

Taaffe, E. J. (ed.) (1970), *Geography*, Prentice-Hall, for the National Academy of Sciences and Social Science Research Council (U.S.).

Tarrant, J. R. (1970), 'Comments on the Use of Trend-Surface Analysis in the Study of Erosion Surfaces', *Transactions*, Institute of British Geographers, 51, 221–2.

Taylor, E. G. R. et al. (1938), 'Discussion of the Geographical Distribution of Industry', *Geographical Journal*, 92, 22–32.

Taylor, P. J. (1974), 'Electoral Districting Algorithms and Their Applications', paper presented to the annual conference of the Institute of British Geographers, University of East Anglia.

(1971), 'Distances within Shapes: An Introduction to a Family of Finite Frequency Distributions', *Geografiska Annaler*, Ser. B, 53, 40–53.

Tobler, W. R. (1967), 'Of Maps and Matrices', *Journal of Regional Science*, 7 (supplement), 275–80.

Törnqvist, G. (1970), 'Contact Systems and Regional Development', *Lund Studies in Geography*, Ser. B, 35.

(1968), 'Flows of Information and the Location of Economic Activities', *Geografiska Annaler*, Ser. B, 50, 99–107.

Törnqvist, G. et al. (1971), 'Multiple Location Analysis', *Lund Studies in Geography*, Ser. C, No. 12, Gleerup.

Townroe, P. M. (1971), 'Industrial Location Decisions. A Study in

Management Behaviour', *Occasional Paper* 15, Centre for Urban and Regional Studies, University of Birmingham.

Toynbee, A. J. (1934–61), *A Study of History*, 12 vols., O.U.P.

Ullman, E. L. (1954), 'Transportation Geography', in P. E. James and C. F. Jones (eds.), *American Geography: Inventory and Prospect*, Syracuse University Press, 311–32.

Unstead, J. F. (1933), 'A System of Regional Geography', *Geography*, XVIII, 175–87.

Ward, B. (1972), *What's Wrong with Economics?*, Macmillan.

Warntz, W. (1973), 'New Geography as General Spatial Systems Theory – Old Social Physics Writ Large?', in R. J. Chorley (ed.) (1973), 89–126.

(1959), *Toward a Geography of Price. A Study in Geo-Econometrics*, University of Pennsylvania Press.

Watson, J. W. (1955), 'Geography – A Discipline in Distance', *Scottish Geographical Magazine*, 71, 1–13.

Weaver, J. C. (1954), 'Crop-Combination Regions in the Middle West', *Geographical Review*, 44, 175–200.

Weber, A. (1909 and 1929), *Theory of the Location of Industries*, University of Chicago Press. Translated from the German by C. J. Friedrich.

Whittlesey, D. (undated), *German Strategy of World Conquest*, Robinson.

Wilkinson, H. R. (1951), *Maps and Politics. A Review of the Ethnographic Cartography of Macedonia*, Liverpool University Press.

Wilson, A. G. (1970), *Entropy in Urban and Regional Modelling*, Pion.

Wooldridge, S. W. (1950), 'Reflections on Regional Geography in Teaching and Research', *Transactions and Papers*, Institute of British Geographers, 16, 1–11.

Wooldridge, S. W., and W. G. East (1966), *The Spirit and Purpose of Geography*, Hutchinson, 3rd edn.

Worswick, G. D. N. (1972), 'Is Progress in Economic Science Possible?', *Economic Journal*, 82, 73–86.

Wrigley, E. A. (1961), *Industrial Growth and Population Change: A Regional Study of the Coalfield Areas of North-West Europe in the Later Nineteenth Century*, C.U.P.

Yeates, M. H. (1969), 'A Note Concerning the Development of a Geographic Model of International Trade', *Geographical Analysis*, 1, 399–404.

(1968), *An Introduction to Quantitative Analysis in Economic Geography*, McGraw-Hill.

(1963), 'Hinterland Delimitation: A Distance Minimizing Approach', *Professional Geographer*, 16, 7–10.
Zipf, G. K. (1949), *Human Behavior and the Principle of Least Effort. An Introduction to Human Ecology*, Addison-Wesley.

Index of Authorities Cited

Subject Index